Advances in Anatomy Embryology and Cell Biology

Vol. 172

Springer
Berlin
Heidelberg
New York
Hong Kong
London
Milan
Paris
Tokyo

M.J. McKinley · R.M. McAllen
P. Davern · M.E. Giles
J. Penschow · N. Sunn
A. Uschakov · B.J. Oldfield

The Sensory Circumventricular Organs of the Mammalian Brain

Subfornical Organ,
OVLT and Area Postrema

With 28 Figures

Springer

Michael J. McKinley, PhD, DSc
Robin M. McAllen, BSc (Hons) PhD, MBChB
Pamela Davern, BSc, Lipt B
Michelle E. Giles, BAppSc
Jenny Penschow, BAppSc, PhD
Nana Sunn, BSc (Hons)
Aaron Uschakov, BSc (Hons)
Brian J. Oldfield, BSc (Hons), PhD

Howard Florey Institute of
Experimental Physiology and Medicine
University of Melbourne, Victoria, 3010, Australia

Mail to: Dr M.J. McKinley
Howard Florey Institute
University of Melbourne, Vic, 3010, Australia
Fax: 61 3 9348 1707
e-mail: mmck@hfi.unimelb.edu.au

ISSN 0301-5556
ISBN 3-540-44367-3 Springer-Verlag Berlin Heidelberg New York

Library of Congress Cataloging-in-Publication Data
The sensory circumventricular organs of the mammalian brain /
M.J. McKinley ... [et al.]
p. cm. – (Advances in anatomy, embryoloy, and cell biology,
ISSN 0301-5556; v. 172)
Includes bibliographical references and index.
ISBN 3-540-00419-X
1. Circumventricular organs. I. McKinley, M.J. (Michael J.), 1946- II. Series.

Springer-Verlag Berlin Heidelberg New York
a member of
BertelsmannSpringer
Science + Business Media GmbH

http://www.springer.de

Typesetting: Stürtz, Würzburg

Printed on acid-free paper 27/3150/ag – 5 4 3 2 1 0

Acknowledgements

The authors wish to acknowledge the support for their research from the National Health and Medical Research Council of Australia, Institute Block Grant 983001; the Ian Potter Foundation; the Robert J. Jr. and Helen C. Kleberg Foundation; and the G. Harold and Leila Y. Mathers Charitable Foundation. We are grateful to Mariella McKinley and Kati Bromley for artwork.

Contents

Abbreviations

ACE	angiotensin converting enzyme
ACTH	adrenocoticotropic hormone
AMPA	alpha-amino-3-hydroxy-5-methylisoxazole-4-propionic acid
ANP	atrial natriuretic peptide
AP	area postrema
AV3V	anteroventral third ventricular wall region
BNP	brain natriuretic peptide
BNST	bed nucleus of the stria terminalis
CCK	cholecystokinin
CGRP	calcitonin gene related peptide
CLIP	coticotropin-like intermediate lobe peptide
CLR	calcitonin-like receptor
CNP	c-type natriuretic peptide
CNS	central nervous system
CSF	cerebrospinal fluid
CTB-gold	cholera toxin b subunit-gold conjugate
CT-HRP	cholera toxin b subunit-horseradish peroxidase conjugate
CVO	circumventricular organ
CVLM	caudal ventrolateral medulla
DNA	deoxyribonucleic acid
Fos-IR	Fos-immunoreactivity
GABA	gamma amino butyric acid
GAD	glutamic acid decarboxylase
GLP-1	glucagon-like peptide-1
HRP	horseradish peroxidase
ICV	intracerebroventricular
IL-1β	interleukin-1 beta
IL-6	interleukin-6
iNOS	inducible nitric oxide synthase
LHRH	luteinising hormone releasing hormone
L-NAME	NG-nitro-L-arginine methyl ester
LPS	lipopolysaccharide
MnPO	median preoptic nucleus
mRNA	messenger ribonucleic acid
MS	melanocyte stimulating hormone
NADPH	nicotinamide adenine di nucleotide phosphate

NMDA	N-methyl-D-aspartic acid
nNOS	neuronal nitric oxide synthase
NO	nitric oxide
NPY	neuropeptide Y
NTS	nucleus of the solitary tract
OVLT	organum vasculosum of the lamina terminalis
PGE_2	prostaglandins E2
PHA-L	phaseolus vulgaris leucoagglutinin
PS	permeability x surface area product
PVN	hypothalamic paraventricular nucleus
PYY	peptide YY
RAMP	receptor-activated modifying protein
RVLM	rostral ventrolateral medulla
SFO	subfornical organ
SOCS-3	suppresor of cytokine signalling-3
SRIF	somatostatin
TLR-4	toll-like receptor 4
TNF-α	tumour necrosis factor alpha
WGA	wheat germ agglutinin
WGA-HRP	wheat germ agglutinin-horseradish peroxidase conjugate

1 Introduction

Circumventricular organs (CVOs) are found in the brains of all vertebrates. These specialised structures are so named because of their proximity to the ventricles of the brain (Hofer 1958). The mammalian brain has fewer CVOs than the brains of other classes of vertebrates; however, it is clear that CVOs still play important functional roles in mammals. The seven CVOs designated in mammals are the subfornical organ, organum vasculosum of the lamina terminalis (OVLT), median eminence, pineal gland, subcommissural organ, area postrema (Fig. 1) and choroid plexuses of the lateral, third and fourth ventricles. Although they are quite diverse in their structures and function, common features are their periventricular midline location in the brain, rich vascularisation, altered blood-brain barrier characteristics and atypical ependymal cells.

Of the seven mammalian CVOs, only three harbour neuronal cell bodies. These are the subfornical organ, OVLT and area postrema, and the neurons within them have efferent neural connections to many other brain regions. Because these three CVOs lack the normal blood-brain barrier, their neurons are exposed to the chemical environment of the general circulation, unlike any other neuronal perikarya in the CNS. Johnson and Gross (1993) termed these three CVOs the sensory circumventricular organs.

The sensory CVOs of the mammalian brain were first described in detail in the early decades of the twentieth century. Wilson (1906a, b) described the morphology of the human area postrema (which he termed the nucleus postremus) in detail, and this initial report came from the University of Sydney, shortly after Australia achieved nationhood. Subsequently, the human subfornical organ, initially termed the intercolumnar tubercle, was described in detail by Tracy Jackson Putnam (1922) working at Johns Hopkins University in Baltimore, although Grafton Elliot Smith, an Australian neuroanatomist and anthropologist who had been a student of Wilson's in Sydney, had initially recognised this structure in the brain of an Australian bat *Nyctophilus timorensis* in 1898. The initial use of the term subfornical organ appears to be by Pines (1926). The last of the three sensory CVOs to be recognised, the organum vasculosum of the lamina terminalis (OVLT), in the brain of the mouse, was first noticed by Gerhard Behnsen in 1927, working at the Anatomical Institute in Kiel in Germany, who observed that the anterior wall of the optic recess was stained by intravitally administered trypan blue.

George B. Wislocki and his co-workers at the Harvard Medical School in Boston, and also Behnsen (1927), made the crucial observation that these regions of the brain, along with the other CVOs, became stained with intravitally administered dyes (Wislocki and Putnam 1920; Wislocki and Putnam 1924; Wislocki and King 1936).

This is unlike the remainder of the brain which excludes intravital dyes because of the existence of a blood-brain barrier. This led to the discovery that these CVOs lacked the blood–brain barrier (Wislocki and King 1936; Wislocki and Leduc 1952). The term 'circumventricular organ' was first used in 1959 by Helmut Hofer, and in the following decade the extensive and specialised vasculature of the CVOs was described for many mammals. In addition, the ultrastructural features of the sensory CVOs, their fenestrated capillary endothelium, neural and glial elements and modified ependyma were described for the first time (see Sect. 4).

From the 1970s onward, the physiological functions of the CVOs became a focus of investigation, particularly after Simpson and Routtenberg (1973) demonstrated that the subfornical organ was the site at which circulating angiotensin II stimulated water drinking, whereas Borison and Brizzee (1951) had earlier proposed that the area postrema was a chemoreceptor trigger zone for vomiting. Extensive physiological investigations of the CVOs, particularly of their role in cardiovascular and body fluid homeostasis, have continued over the past 30 years. In the 1980s, the advent of modern neural tracing techniques enabled the direct neural connectivity of the CVOs to be mapped (see Sect. 6).

Another aspect of the CVOs that has blossomed during the past 20 years is our understanding of the neurochemistry of these regions. The development of immunohistochemistry, in vitro autoradiographic receptor binding techniques and in situ hybridisation histochemistry has enabled mapping of neurotransmitters and neuropeptides in CVOs, as well as their receptors and mRNA. The contributions to this area are described in Sect. 5. During the past two decades, electrophysiological and ion channel studies have also greatly added to our knowledge of the sensory CVOs (see Sect. 8).

The most recent methodological advance that has significantly increased our ability to understand the sensory CVOs is the utilisation of *c-fos* expression as an indicator of increased neuronal activity to map populations of activated neurons and pathways throughout the central nervous system (Morgan et al. 1987; Hunt et al. 1987) (see Sect. 7).

This monograph reviews these various aspects of the sensory CVOs. Our major focus has been the regulation of body fluids and the cardiovascular system, and our treatment of these issues probably reflects such interests. However, we are mindful that the CVOs have other roles as well, in particular fever and emesis. An aim of this review is to correlate structure with function in considering the morphology, neural connectivity and neurochemistry of the sensory CVOs. We have built on the foundations of earlier excellent reviews of various aspects of CVO structure and function and have attempted to update and extend them. These include the reviews of Hofer (1969), Weindl (1973), Leonhardt (1980), Johnson and Gross (1993), those in the three-volume work *Circumventricular Organs and Body Fluids* edited by Gross (1987), the volumes that resulted from the Rheinhardsbrunn Conferences—the last of these appearing as a volume of *Progress in Brain Research* in 1992, and the review by Ferguson and Bains (1996) of CVO electrophysiology. In addition, several reviews specifically on the subfornical organ (Dellman and Simpson 1979; Dellman 1998) and the area postrema (Borison 1989; Leslie 1986; Miller and Leslie 1994) provide a specialised background to this review.

2 Location, General Structure and Ependymal Cells of Sensory Circumventricular Organs

Located at seemingly strategic points in the brain's ventricular system (Figs. 1, 2), the sensory CVOs are midline structures situated in the walls of the third and fourth ventricles. The subfornical organ and OVLT occupy the dorsal and ventral extremities of the lamina terminalis, which embryologically is the most anterior aspect of the developing central nervous system (Bayer and Altman 1987). The subfornical organ appears to guard the entrance of the third ventricle, at the point where the cerebrospinal fluid (CSF) from the lateral ventricles enters the third ventricle, and where its CSF begins to bathe the ventricular surface of the diencephalon and brain stem. At the posterior extremity of the brain, the area postrema, situated in walls of the fourth ventricle at the caudal end of the medulla oblongata, appears to guard the entrance to the central canal of the spinal cord.

To avoid confusion for the reader, at the outset of this monograph it should be pointed out that each of the sensory CVOs has been referred to in the literature by a number of various names and acronyms and these are given in Table 1. Throughout this work, we will employ the most commonly used terminology at present: subfornical organ, abbreviated to SFO in our diagrams and figures; organum vasculosum of the lamina terminalis, which will be abbreviated to OVLT throughout the text and figures; and area postrema, abbreviated to AP in figures.

2.1 Subfornical Organ

2.1.1 Location

The subfornical organ is a small, ovoid structure that bulges into the midline anterior wall of the third ventricle (Figs. 2, 3) dorsal to the anterior commissure at the confluence of the interventricular foramina of Monro in mammals (Putnam 1922; Akert et al. 1961). This sensory CVO lies between the columns of the fornix and is attached to the hippocampal commissure dorsally. It has a ventral stalk which merges with the median preoptic nucleus, and its dorsal crest joins the tela choroidea of the third ventricle (Fig. 3a). The subfornical organ is somewhat translucent in appearance and is richly vascularised in all species studied (Hofer 1958, Spoerri 1963; Duvernoy and Koritke 1964; Duvernoy and Koritke 1965; Duvernoy and Koritke 1969; Dellman 1987). In many species (e.g. rat, rabbit, sheep), the subfornical organ is quite prominent in coronal sections as it extends into the lumen of the third ventricle and labels

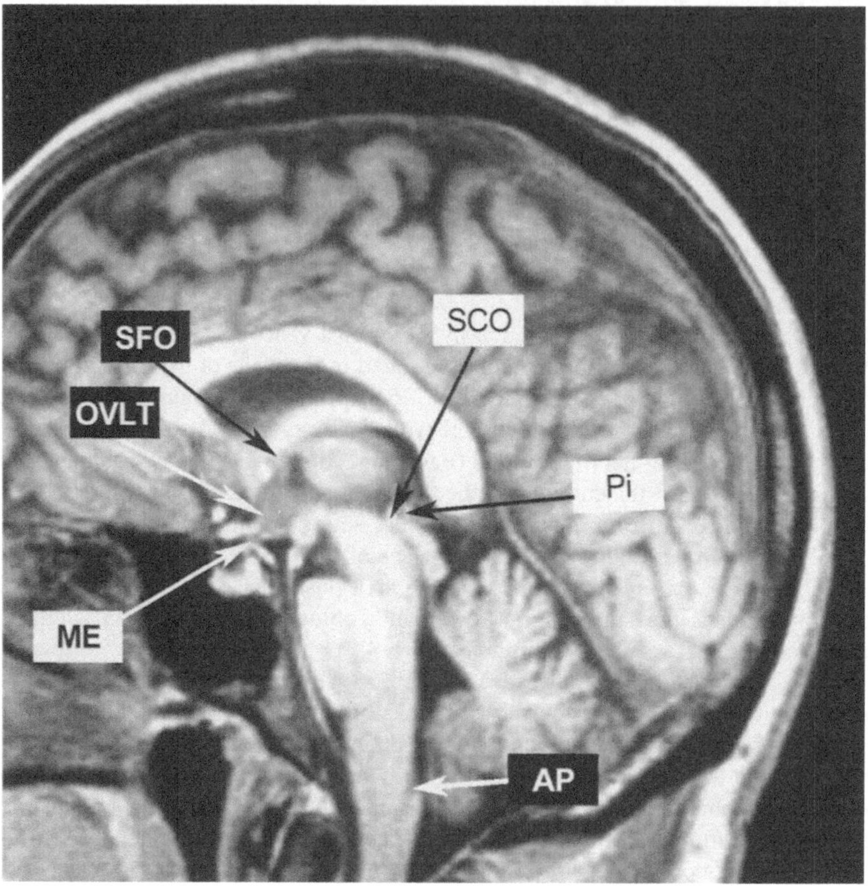

Fig. 1 The position of the circumventricular organs in the human brain as shown by magnetic resonance imaging. The three sensory circumventricular organs, the subfornical organ, OVLT and area postrema are indicated by *black labelling*. The other circumventricular organs shown are the median eminence, pineal gland and subcommissural organ, although this latter structure regresses shortly after birth in the human brain. The other circumventricular organ (not shown) is the choroid plexus found in the lateral, third and fourth ventricles. *AP*, area postrema; *ME*, median eminence; *OVLT*, organum vasculosum of the lamina terminalis; *Pi*, pineal gland; *SCO*, subcommissural organ; *SFO*, subfornical organ

Table 1 Synonyms used for the sensory circumventricular organs

Subfornical organ	OVLT	Area postrema
Intercolumnar tubercle	Supraoptic crest	Nucleus postremus
Ganglion psalterii	Preoptic organ	
Interventricular organ	Terminal plate organ	
	Dorsal chiasmatic nucleus	
	Optic recess organ	
Medial (vascular) prechiasmatic gland		

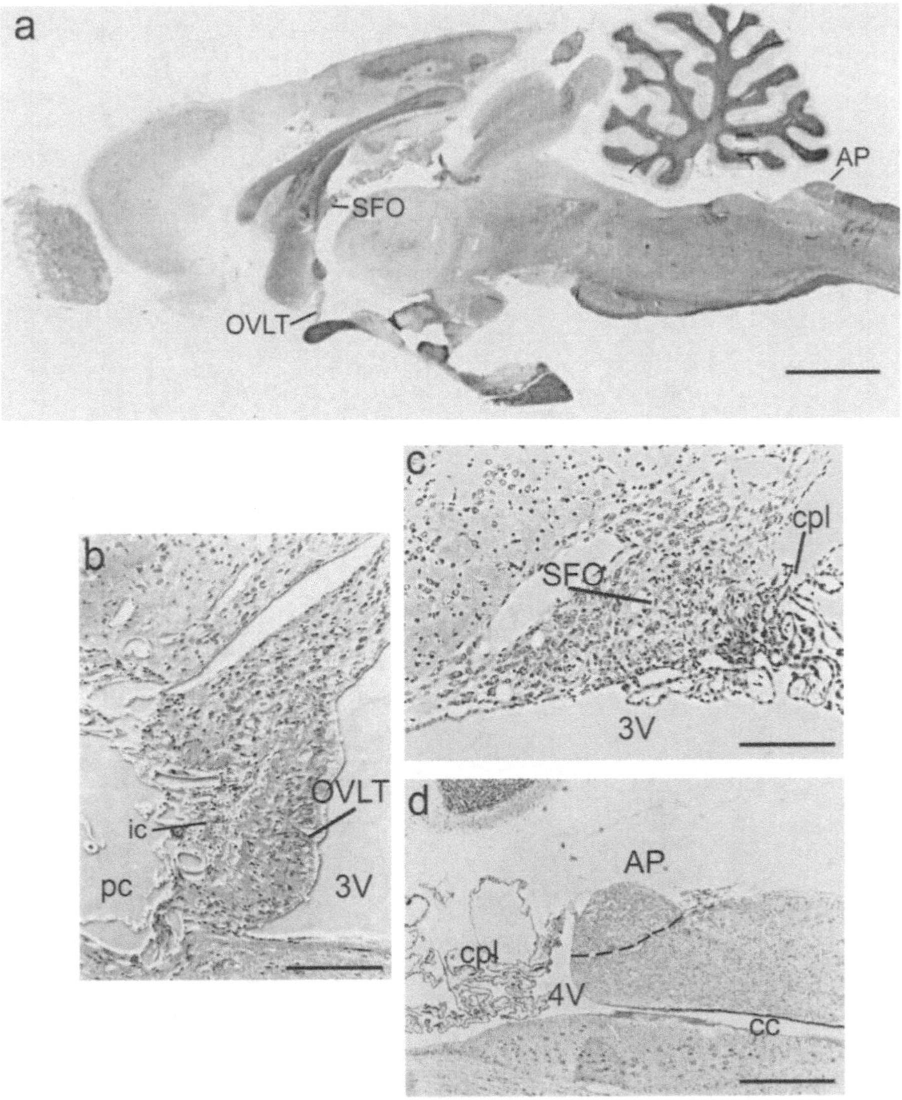

Fig. 2a–d Mid-sagittal sections of the rat brain showing the sensory CVOs. **a** Low-power section of the rat brain. **b** The OVLT. **c** The subfornical organ. **d** The area postrema (its ventral boundary is indicated by the *interrupted line*). *AP*, area postrema; *ic*, intrapial capillary plexus; *cpl*, choroid plexus; *cc*, central canal; *OVLT*, organum vasculosum of the lamina terminalis; *pc*, prechiasmatic cistern; *SFO*, subfornical organ; *3V*, third ventricle; *4V*, fourth ventricle. Stain, cresyl fast violet. *Calibration bars*=2.5 mm in **a**, 0.25 mm in **b** and **c**, and 0.75 mm in **d**

strongly with Nissl stains (McKinley et al. 1983b; Mendelsohn et al. 1988); however, in the human brain, the subfornical organ has been described as miniscule (Mark and Farmer 1984). It can be seen in the wall of the third ventricle (Fig. 3a, b, e) as a slight eminence just below the tela choroidea of the choroid plexus of the third ven-

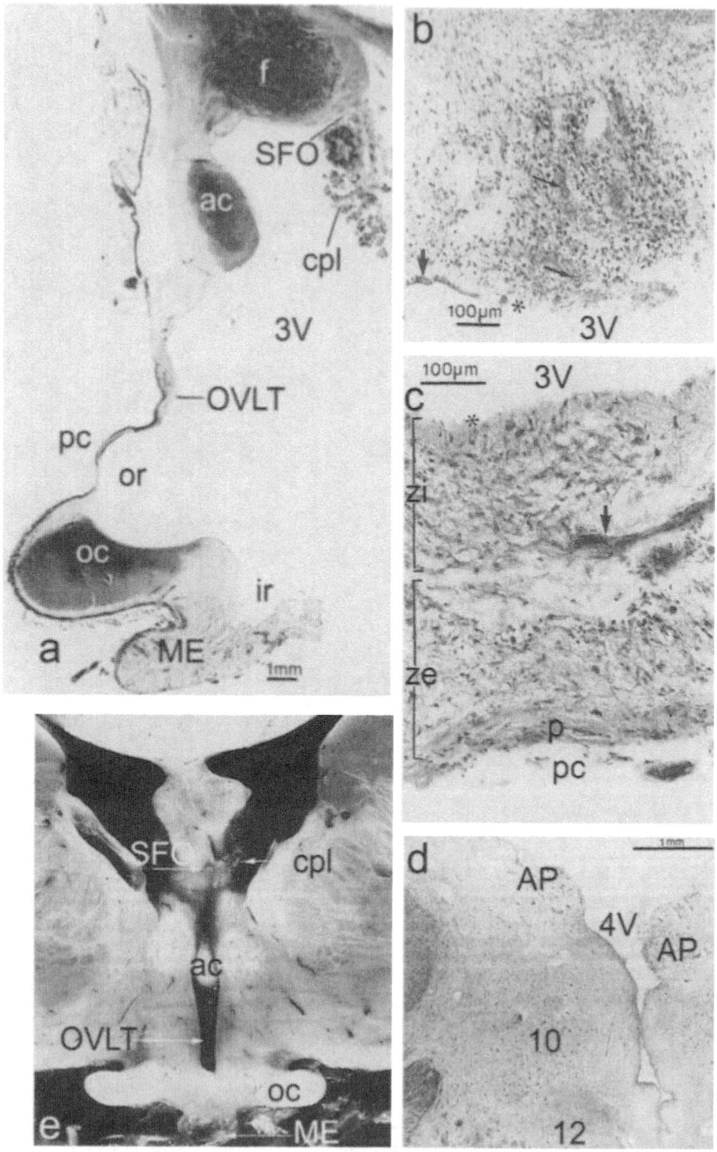

Fig. 3a–e Sensory CVOs of the human brain. **a** Mid-sagittal section through the anterior wall of the third ventricle showing the subfornical organ, OVLT and median eminence. **b** Coronal section through the human subfornical organ. **c** Horizontal section through the OVLT. *Asterisk* indicates flattened ependyma; *arrow* indicates a blood vessel. **d** Coronal section through the human area postrema. **e** Location of the subfornical organ and OVLT as seen in post-mortem tissue cut through the anterior third ventricle. *ac*, Anterior commissure; *AP*, area postrema; *cpl*, choroid plexus; *ir*, infundibular recess; *ME*, median eminence; *oc*, optic chiasm; *or*, optic recess; *OVLT*, organum vasculosum of the lamina terminalis; *p*, pia mater; *pc*, prechiasmatic cistern; *SFO*, subfornical organ; *ze*; external zone; *zi*, internal zone; *3V*, third ventricle; *4V*, fourth ventricle; *10*, dorsal motor nucleus of the vagus; *12*, hypoglossal nucleus. (Reproduced with permission from McKinley and Oldfield 1991)

tricle at the junction of the two fornical columns (Duvernoy and Koritke 1969). In the marsupial brain, which lacks a corpus callosum, it is set more rostrally over the large anterior commissure (Akert et al. 1961).

2.1.2
Ependymal Surface of the Subfornical Organ

The ependymal cells that line the ventricular surface of the subfornical organ differ considerably from the surrounding ependyma and, in fact, differ considerably across species. In general, ependymal cells of the subfornical organ are flattened, lack the copious cilia of normal ependymal cells and have tight junctions between adjacent cells, providing an obstacle to diffusion of substances between the parenchyma of the subfornical organ and the third ventricle (Dellman 1987). In the rat, for example, several scanning electron microscope studies show that the ependymal elements vary across the ventricular surface of the organ. In the central region of the subfornical organ's surface, the ependymal cells are hexagonal and lacking in cilia or microvilli (Phillips et al. 1974; Dellman 1987, 1998). Around this central region is an intermediate annular zone, where convex ependymal cells may exhibit either microvilli or clumps of cilia. Some ependymal cells may exhibit a single cilium, whereas the most lateral region of the rat's subfornical organ shows ependymal cells with many surface cilia (Dellman 1987). The ventricular surface of the subfornical organ of the rabbit has the appearance of a mosaic of regularly shaped hexagonal cells (Fig. 4), some of which exhibit a single cilium and microvilli (Weindl and Joynt 1972). The subfornical organ of the cat shares these features (Pfenninger 1969; Felix et al. 1977), whereas the ependymal surfaces of the subfornical organs of the sheep (Fig. 5) and goat are quite different. In these species, the subfornical organ exhibits many intertwining processes and supraependymal cells, possibly macrophages and neurons, that are situated on these processes.

2.2
OVLT

2.2.1
Location and Structure

The OVLT is situated in the midline anterior wall of the third cerebral ventricle immediately dorsal to the optic chiasm (Figs. 2a, 3a). It extends towards the anterior commissure, reaching more or less halfway to this major fibre pathway, depending on the species. Rostrally, the OVLT is bounded by the pial surface of the CSF-filled pre-chiasmatic cistern. Large blood vessels within the pia mater give rise to a dense intrapial capillary plexus that invaginates into the body of the OVLT rostrally, while caudally, the bulbous base of the OVLT protrudes into the optic recess of the third ventricle. This is particularly pronounced in some species such as the dog, rabbit and sheep (Figs. 6, 7, 8). The shape and general structure of the OVLTs of many mammals, including several primates, has been described in the literature (Behnsen

Fig. 4a, b Scanning electron micrographs of the ventricular surface of the subfornical organ of the rabbit. **a** Low-power view from the dorsal caudal aspect. *Arrowheads* indicate the ventral boundary of the subfornical organ. ×120. **b** High-power view of the ependymal surface of the subfornical organ. *Arrow* indicates microvilli on the surface of a non-ciliated ependymal cell; *double arrowhead*, single cilium. ×1,350. *ce*, Ciliated ependyma; *SFO*, subfornical organ

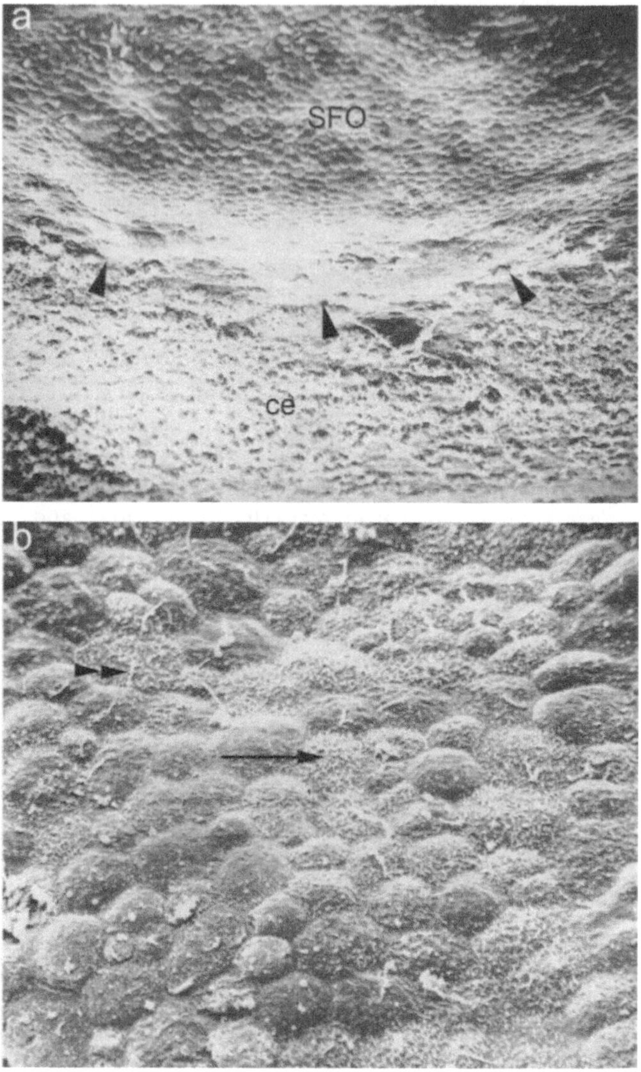

1927; Hofer 1958; Mergner 1961; Duvernoy and Koritke 1964; Landas and Phillips 1987; Weindl and Joynt 1972; McKinley et al. 1983a; McKinley et al. 1987b). The body of the OVLT narrows in the dorsal direction, giving it the shape of an elongated, arching structure or crest, and this is undoubtedly the reason for its early designation as the supraoptic crest. Rostral and lateral to the OVLT is the nucleus of the diagonal band of Broca, the medial preoptic region is adjacent to much of its lateral aspect, and the anteroventral periventricular nucleus bounds its most caudo-lateral aspect.

Just dorsal to the OVLT is the median preoptic nucleus (MnPO), and there has been some uncertainty regarding the boundary between the OVLT and MnPO in the rat and other species. We define the boundary of the dorsal border of the OVLT in

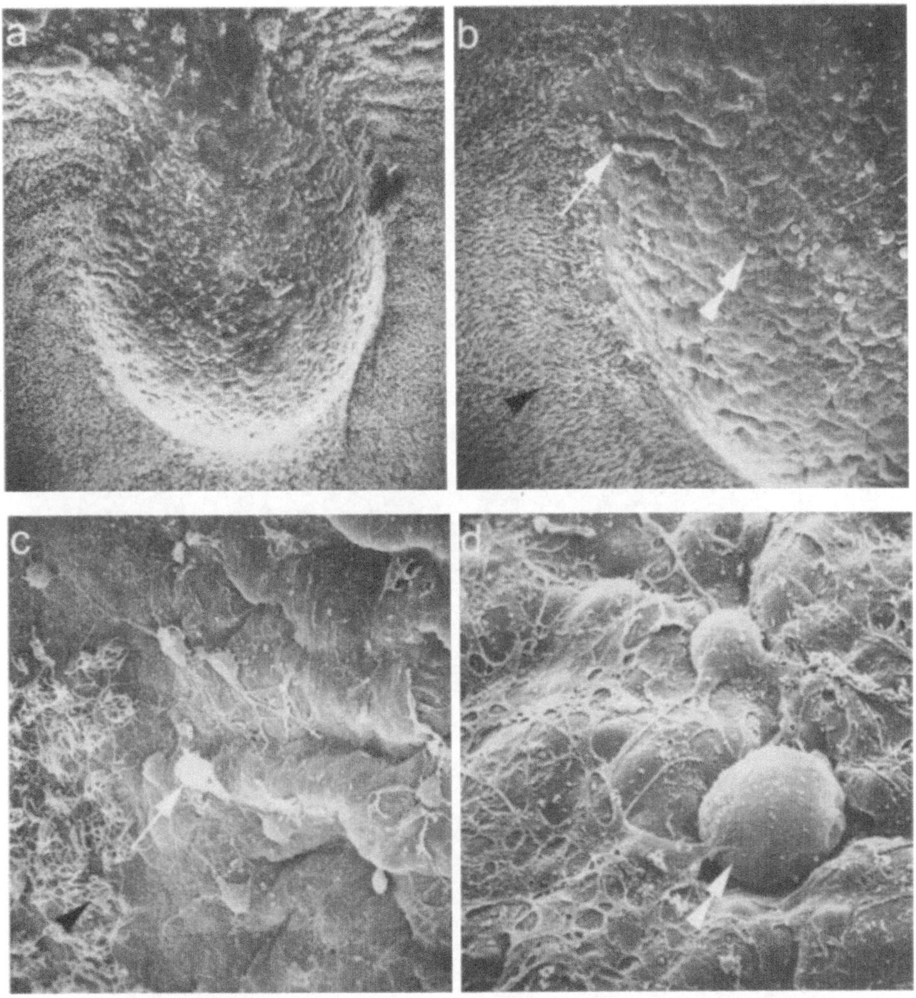

Fig. 5a–d Scanning electron micrographs of the ventricular surface of the subfornical organ of the sheep. **a** ×40. **b** ×80. **c** ×250. Ciliated ependyma (*black arrowhead*) is seen lateral to the subfornical organ, and the same supraependymal cell is shown by an *arrow* in **b** and **c**. **d** Many flattened cellular processes are seen covering the surface of the subfornical organ. The same supraependymal cell is indicated by *double arrowhead* in **b** and **d**. ×600

the rat by using a number of criteria, including the absence of a blood-brain barrier, neurochemical characteristics, cytoarchitectonic features and function reflected by immediate-early gene expression. With these criteria, the boundaries of the OVLT can be seen in Figs. 11 and 12, Sect. 3. The arching group of cells seen in the dorsal part of the OVLT, which we refer to as the dorsal cap of the OVLT, is exposed to the haemal milieu as shown by the passage of blood-borne horseradish peroxidase (HRP) into its interstitium (Bisley et al. 1996). The blood–brain barrier is intact in the median preoptic nucleus, and because cells of the dorsal cap are in a region with no blood–brain barrier, we consider these to be part of the OVLT. In addition, the

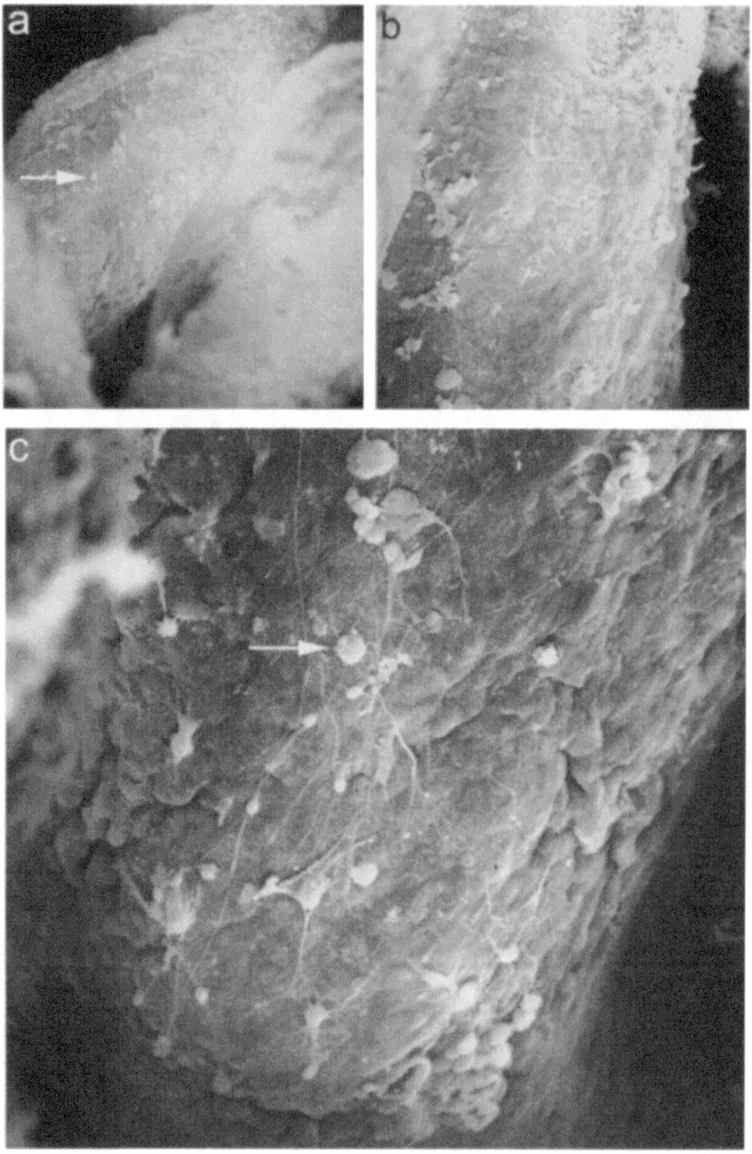

Fig. 6a–c Scanning electron micrographs of the ventricular surface of the OVLT of the dog. **a** The OVLT is viewed from a caudal ventrolateral aspect. ×80. **b** ×96. **c** ×170. The same supraependymal neuron is indicated by an *arrow* in **a** and **c**. Ependymal cells lack cilia

cells in this arching dorsal cap region exhibit calretinin immunoreactivity, whereas the median preoptic nucleus does not (see Fig. 12). In regard to functional aspects of the CVOs, high-affinity receptor binding sites for the hormone relaxin have been identified in the OVLT and we have observed intense expression of the immediate-early gene *c-fos* in the dorsal cap region (Fig. 9) but not in the median preoptic nu-

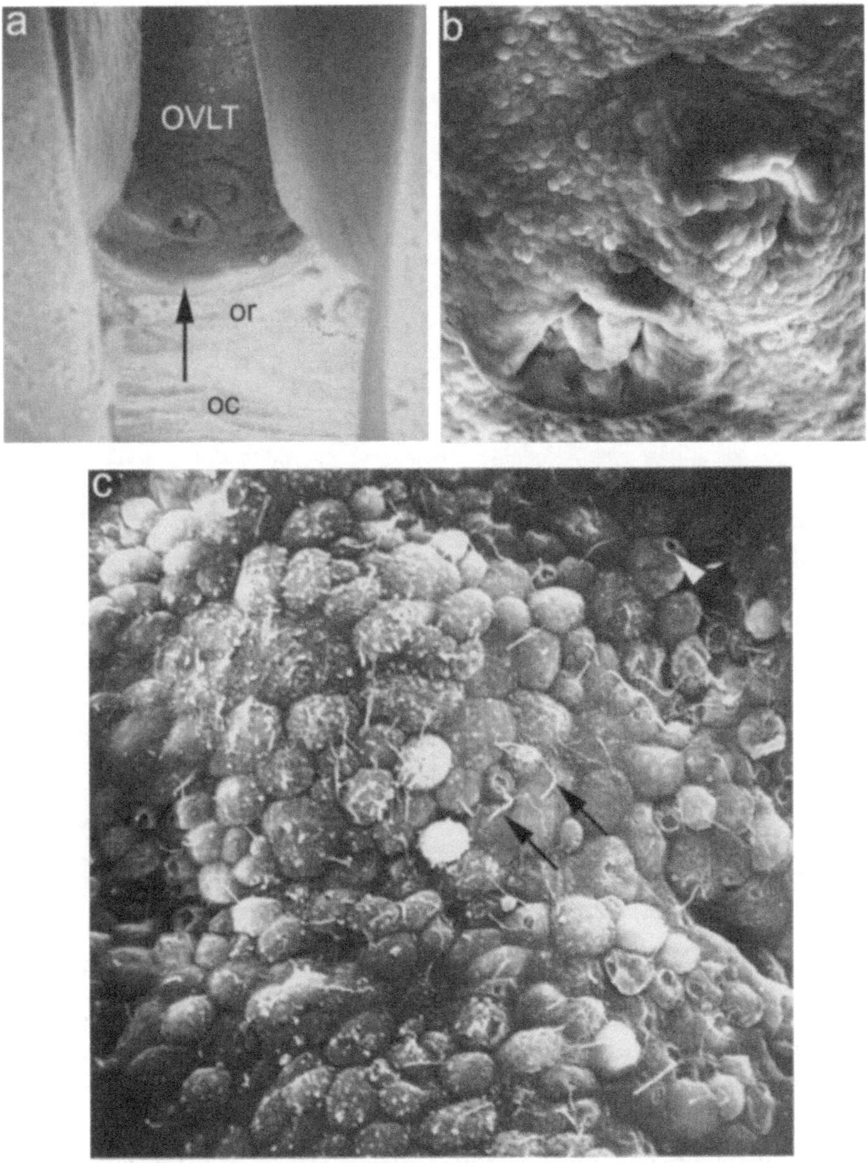

Fig. 7a–c Scanning electron micrographs of the ventricular surface of the OVLT of the rabbit. **a** The OVLT is viewed from a caudal aspect, and its most ventral aspect is indicated by the *black arrow.* ×28. **b** ×155. **c.** Ependymal cells exhibit a single cilium *(black arrows)*, and a surface eruption is indicated by the *white arrowhead.* Most ependymal cells have blunt protrusions on their surface. ×560. *oc*, Optic chiasm; *or*, optic recess; *OVLT*, organum vasculosum of the lamina terminalis

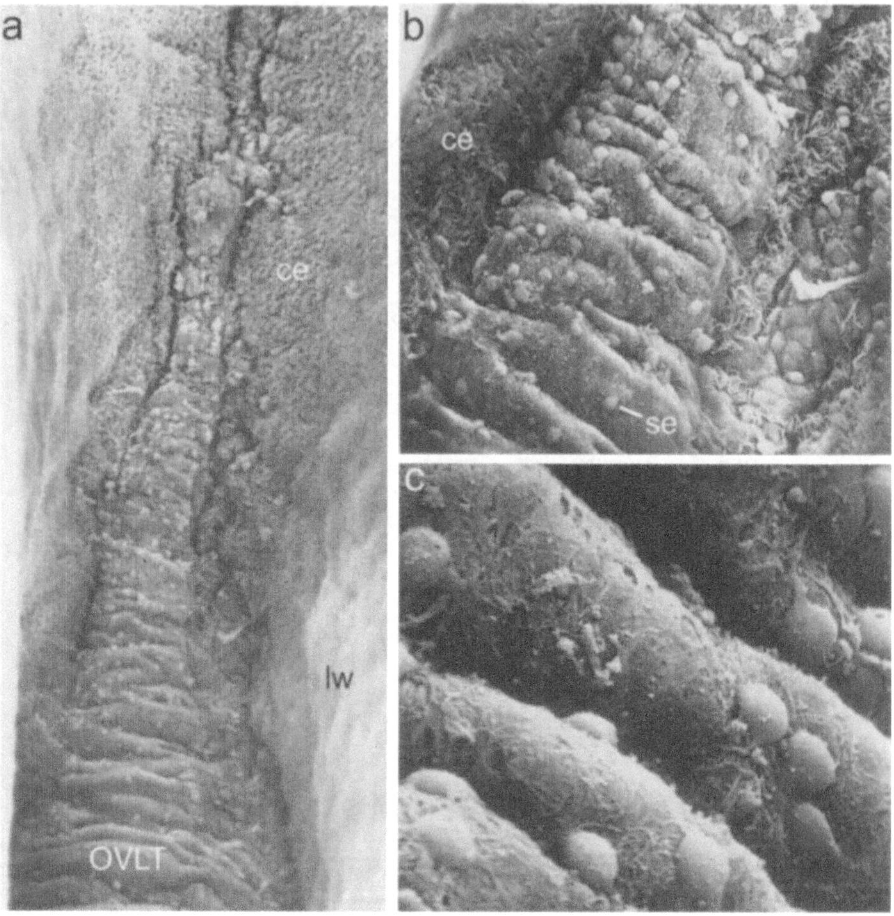

Fig. 8a–c Scanning electron micrographs of the ventricular surface of the dorsal part of the OVLT of a sheep. **a** The OVLT is viewed from a caudal aspect, ×60. **b** ×120. **c** Higher magnification of the surface of the sheep OVLT showing many interfolding processes and supraependymal cells. ×350. *ce*, Ciliated ependyma; *lw*, lateral wall of the third ventricle; *se*, supraependymal cell; *OVLT*, organum vasculosum of the lamina terminalis

cleus, in response to intravenously infused relaxin (McKinley et al. 1998a). When relaxin was infused intracerebroventricularly (i.e. on the brain side of the blood–brain barrier), by contrast, it did stimulate Fos expression in both the OVLT and MnPO (McKinley et al. 1997). These data show that circulating relaxin may gain entry to its receptors in the dorsal cap region of the OVLT, but it does not gain access to the MnPO behind the blood–brain barrier, providing another reason for defining the arching group of neurons in the dorsal cap as part of the OVLT. In Fig. 9, the lateral boundaries of the OVLT are marked by neurons which usually express Fos when the blood levels of angiotensin II increase and are usually stained by NADPH diaphorase (see Fig. 20a, Sect. 5).

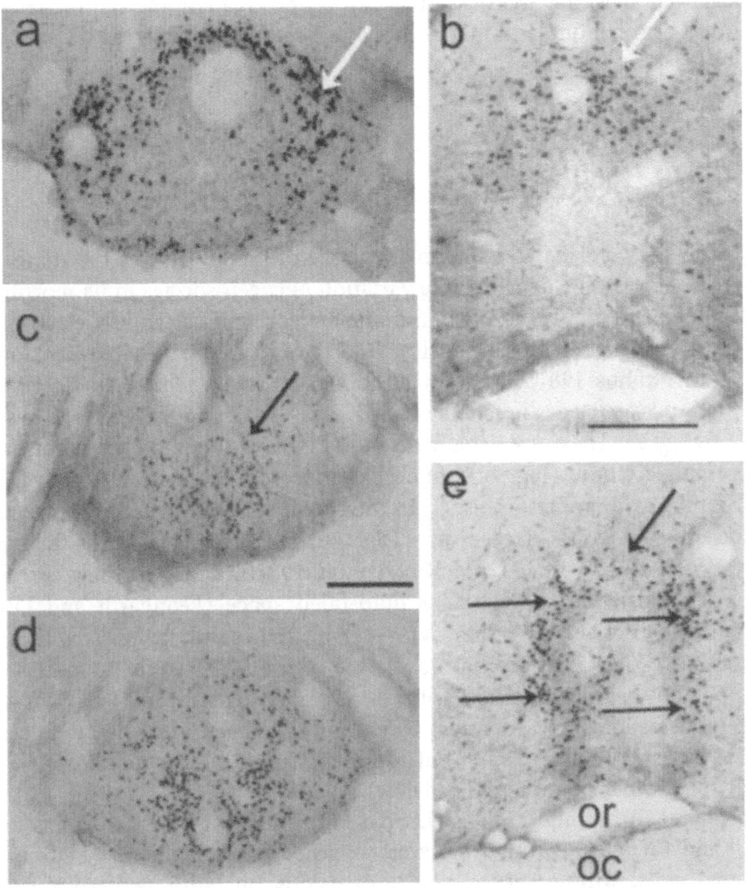

Fig. 9a–e Coronal sections of the subfornical organ and OVLT of rats showing Fos-like immunoreactivity (Fos-LI, seen as *black dots*) in different subgroups of neurons within these circumventricular organs. **a** Subfornical organ of a rat that had been intravenously infused with relaxin at 25 μg/h, showing Fos-LI in the outer shell (*arrow*). **b** OVLT of a rat that had been intravenously infused with relaxin at 25 μg/h, showing Fos-LI (*arrow*) in the dorsal cap region. **c** Subfornical organ of a rat that had been intravenously infused with angiotensin II at 0.3 μg/h, showing Fos-LI in its ventromedial core (*arrow*). **d** Subfornical organ of a rat that had been intravenously infused with angiotensin II at 1 μg/h, showing Fos-LI throughout this circumventricular organ. **e** OVLT of a rat that had been intravenously infused with angiotensin II at 1 μg/h, showing Fos-LI in the lateral part of the OVLT (*small arrows*). Very few cells with Fos-LI are observed in the dorsal cap region (*thick arrow*). *oc*, Optic chiasm; *or*, optic recess of the third ventricle. *Calibration bars*=100 μm

Immediately over the arching dorsal cap of the OVLT in the rat, the lateral wings of the MnPO spread out, and may sometimes be confused with the dorsal cap region of the OVLT.

2.2.2
Ependymal Surface of the OVLT

In regard to the ventricular ependymal surface of the OVLT, like the subfornical organ there appears to be considerable variation between species in its morphology. In general, the ependyma is modified and mostly lacks cilia. Scanning electron microscopy has shown that in species such as the rat and rabbit (Weindl and Joynt 1972; Landas and Phillips 1987; McKinley et al. 1987b; Yamaguchi et al. 1993), epithelial cells with convex surfaces in 'cobblestoned' array, make up the ventricular surface of the OVLT (Fig. 7). These are usually devoid of cilia, although in the rabbit some may exhibit a single cilium (Fig. 7c), much like those observed in its subfornical organ (Fig. 4b); microvilli are also present. In the rabbit some ventricular protrusions, furrows and unusual 'crater-like' features (Fig. 7b) are observed towards the base of the OVLT (Weindl and Joynt 1972; McKinley et al. 1987b) and these may correspond to subependymal lakes of perivascular fluid-filled space (Leonhardt and Lindemann 1973; Yamaguchi et al. 1993). Tanycytes—elongated ependymal cells with long basal processes—are also observed on the surface of the OVLT in these species and may have important barrier functions with the surrounding neuropil (see Sect. 4). In the rat and mouse some supraependymal neurons have been identified (Scott and Paull 1983; Landas and Phillips 1987), but these appear to be less numerous than in the OVLT of some other mammals (see below).

In primates, including humans (Landas and Phillips 1987) and squirrel monkeys (Weindl and Joynt 1972), the ependyma consists of flattened, elongated non-ciliated ependymal cells, without ventricular protrusions but with numerous neuron-like supraependymal cells. These cells have many fine processes and neurites crossing the surface of the OVLT. Their function is unknown. In the OVLT of the dog, many neuronal-type supraependymal cells are also seen with fine branching processes (Fig. 6c) and non-ciliated, flattened ependyma (Landas and Phillips 1987; McKinley et al. 1987b).

The appearance of the ventricular surface of the ruminant OVLT may be quite different to those described above. In sheep (Fig. 8), the ventricular surface is composed of myriads of interfolding and entwining flattened cellular processes, on which supraependymal cells abound (McKinley et al. 1983a). A similar morphology is present on the ventricular surface of the goat OVLT, and in both species many deep horizontal furrows regularly cross the ventricular surface from its base to its most dorsal tip. In addition to the lamella of flattened processes, fine, long fibres also radiate from some surface cells and appear to penetrate into the body of the OVLT. The precise cell type has not been identified yet, but many have the appearance of phagocytes whereas others resemble neurons.

Fig. 10. a Scanning electron micrographs of the ventricular surface of the area postrema of a sheep. A supra-ependymal neuron is indicated by the *arrow*. x250. **b** Ventricular surface of the area postrema of the dog. ×320

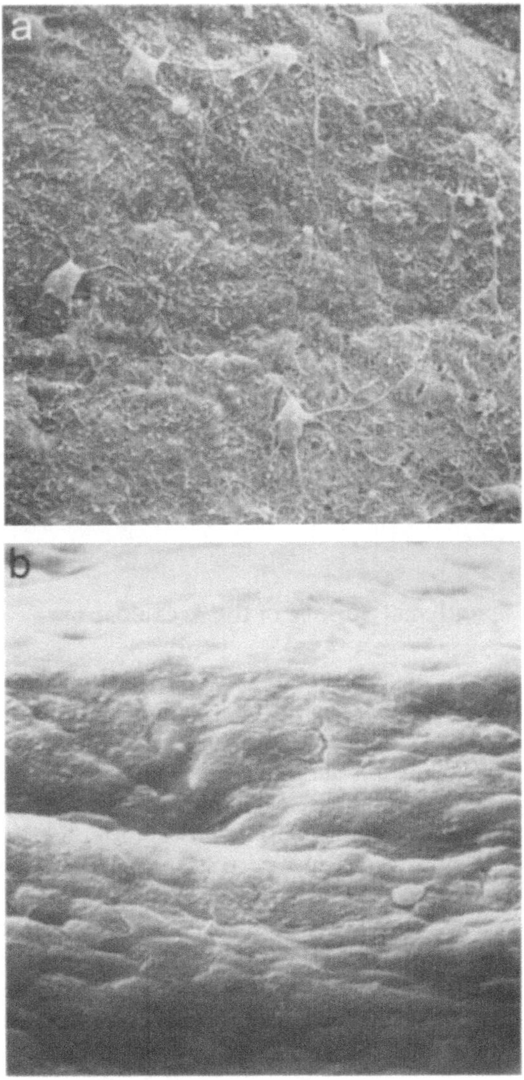

2.3 Area Postrema

2.3.1 Location and Structure

The most caudal of the sensory CVOs, and the first to be recognised, the area postrema was named by Retzius in 1896 and described in detail by Wilson (1906a, b) 10 years later. It is situated in the wall of the fourth ventricle, near the opening of the central canal (Figs. 2, 3d). Its appearance varies amongst species so that in larger mammals such as humans (Fig. 3d), monkeys, dogs and ruminants it is bilaterally represented in the medulla oblongata, in the ventrolateral wall of the fourth ventricle, ventral to the choroid plexus. In these species, it bulges into the ventricular lumen, converging caudally to form the roof of the central canal. The immediately adjacent medullary region to the area postrema is the subpostremal part of the medial subnucleus of the nucleus of the solitary tract (NTS), which lies lateral and ventral to the CVO. The tela choroidea is attached at the dorsolateral margin of the area postrema. In rodents and lagomorphs (e.g. rat, rabbit, mouse), the area postrema is a midline hemisphere at the level of the obex (Fig. 2d), whereas the ferret has a prominent midline hump with lateral extensions rostrally (Reynolds et al. 1991). In both instances, it overlies the posterior part of the NTS, which together with the dorsal motor nucleus of the vagus nerve comprises the dorsal vagal complex. A glial-rich region, the funiculus separans, separates the area postrema from the NTS. Similar to the other CVOs, the area postrema is richly vascularised, being penetrated by many arterioles and capillaries which give it a sponge-like appearance in histological sections. Rostrally it is connected to the tela choroidea of the choroid plexus.

2.3.2 Ependymal Surface of the Area Postrema

The ependymal surface of the area postrema has been studied with the scanning electron microscope in the brain of humans (Leslie 1986), rats (Torack and Finke 1971), rabbits (Weindl and Joynt 1972) and opossums (Scott and Paull 1983). Like the other CVOs, the area postrema has a modified ependyma. The ependymal cells are flattened and non-ciliated and often exhibit microvilli. In the rabbit, the surface ependymal cells resemble those of the subfornical organ and OVLT in that species and have numerous ventricular protrusions on the cell surface extending into the ventricular lumen (Weindl and Joynt 1972). In Fig. 10a, which shows the surface of the area postrema of the sheep, many processes can be seen covering its surface and supraependymal cells with long narrow unbranching processes are evident. The surface is devoid of cilia, in contrast to the adjacent medullary surface. The ependymal surface of the dog's area postrema is composed of non-ciliated, cobblestoned ependymal cells, without supraependymal neurons (Fig. 10b).

3 Regional Subdivisions Within Sensory Circumventricular Organs

3.1 Subfornical Organ

It is clear from several points of view that the subfornical organ is not a uniform structure. The subfornical organs of several mammalian species were described by Akert et al. (1961) and subdivided into three regions: a dorsal stalk, the body and ventral stalk. Dellmann and Simpson (1976) also recognised three divisions within the subfornical organ, and, more recently, Sposito and Gross (1987) and Shaver et al. (1990) subdivided the subfornical organ of the rat into a number of regions, largely on the basis of capillary morphology and distribution. These were rostral, transitional, central and caudal subregions, with dorsal, ventromedial and lateral zones of the latter three subregions. Other studies of neural connectivity, receptor binding, histochemical staining and functional neuroanatomy utilising *c-fos* expression are indicative of two major functional subdivisions of the rat subfornical organ: the first is a rostrodorsal 'outer shell' which includes the ventral and dorsal stalks of the original classification and dorsal and lateral zones of all regions specified by Sposito and Gross (1987); the second subdivision is a central 'ventromedial core' of the subfornical organ, corresponding to the ventromedial zone of Sposito and Gross (1987). A schematic diagram of the subdivisions of the rat subfornical organ is presented in Fig. 11.

Our simplification of this classification into two functional zones in the rat is based largely on the following morphological, neuroanatomical and functional criteria.

1. The density of capillariesis greatest in the ventromedial core of the subfornical organ(Sposito and Gross 1987). The presence of fenestrationsin the capillary endothelium, indicative of a lack of a blood-brain barrier, is confined to the ventromedial coreregion (Krisch et al. 1978).
2. The appearance of the ependyma of the ventromedial coreregion is different to the remainder of the subfornical organ; in particular, there is a total lack of ciliathere (see Sect. 2).
3. The great majority of the known efferent neural connections, particularly the major ones (e.g. to supraoptic, paraventricular, median preoptic and lateral hypothalamic nuclei), arise from neurons in the outer shell of the subfornical organ. The main neural pathway that originates in the ventromedial core of the subfornical organis a projection to the rostral bed nucleus of the stria terminalis(Swanson et al. 1986) (see Sect. 6).

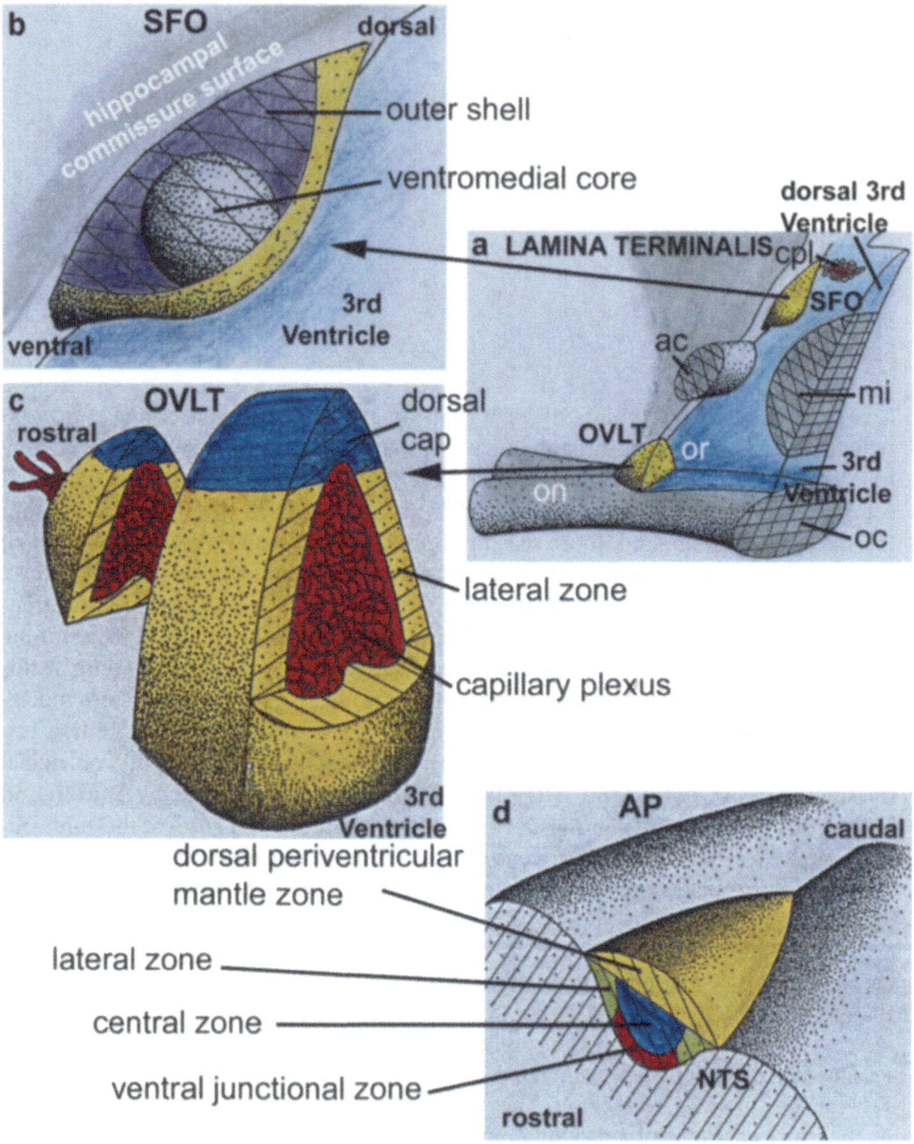

Fig. 11a–d Schematic representation of the functional subdivisions proposed for sensory circumventricular organs. **a** Diagram of the lamina terminalis to show the position of the subfornical organ and OVLT in the anterior wall of the third ventricle. **b** Diagram of the subfornical organ. **c** Diagram of the OVLT. **d** The subdivisions of the area postrema as proposed by Gross (1991). These diagrams are not drawn to scale or proportion. *Parallel or hatched lines* indicate cut faces. *ac*, Anterior commissure; *AP*, area postrema; *cpl*, choroid plexus; *mi*, massa intermedia; *NTS*, nucleus of the solitary tract; *oc*, optic chiasm; *on*, optic nerve, *or*, optic recess; *OVLT*, organum vasculosum of the lamina terminalis; *SFO*, subfornical organ

4. The effects of blood-borne angiotensin IIshow regional differences within the subfornical organ. Using the rate of glucose utilisation as measured by the uptake of radiolabelled 2-deoxyglucose, Shaver et al. (1990) observed that metabolic activity increased in the ventromedial corein response to intravenous infusion of angiotensin II. However, the rate of infusion rate of angiotensin II in this study (2.5 μg/min) was extremely high and much greater than rates that produce blood levels within physiological or pathophysiological ranges. We infused angiotensin II intravenously to ratsat approximately 0.02 μg/min and observed stimulation of neurons throughout the subfornical organ as indicated by the increase in *c-fos*expression; however, there was a greater preponderance of labelled neurons in the ventromedial core (McKinley et al. 1992a). Later, we made intravenous infusion of angiotensin II at even lower infusion rates (0.005 μg/min) and observed that Fosexpression was almost exclusively in the ventromedial core (Fig. 9c), indicating that this division of the subfornical organis the most sensitive to angiotensin II, or most readily accessed. Although angiotensin II AT_1receptorsare located throughout the subfornical organ, there is a higher density in its ventromedial core as seen with in situ hybridization, in vitro autoradiographic binding and immunohistochemistry (Lenkei et al. 1995, 1997; McKinley et al. 1998; Giles et al. 1999). This probably explains why Fos expression in response to relatively low intravenous doses of angiotensin II is predominantly in the ventromedial core. Angiotensin-immunoreactive fibres also terminate preferentially in the ventromedial core (Lind et al. 1984; Lind 1987).
5. Immunohistochemical labelling of calbindin-containing neurons is exclusively in the ventromedial core(Fig. 12d), whereas that for calretininis totally absent from this core and confined to the outer shell(Fig. 13b, c).
6. Cholinergic fibres, as detected by cholinesterasestaining, are localised to the outer shell(Lind 1987), and intracerebroventricularly administered carbacholstimulates expression of *c-fos*in the outer shell.
7. Immunoreactivity for a serotonin transporteris predominantly in the ventromedial core, with fewer such serotonergic nerve terminals in the outer shell (see Fig. 21b, Sect. 5).
8. Fosexpression in response to systemic hypertonicityis observed predominantly in the outer shell(McKinley et al. 1992c).
9. Intravenous or intracerebroventricularly administered relaxin, a hormone secreted during pregnancy, causes intense expression of Fosin many neurons of the outer shell(Fig. 9) of the subfornical organbut not in the ventromedial core (McKinley et al. 1997a, 1998).

Although this delineation of an outer shell and inner core has been made for the rat subfornical organ, it is possible that such a division may also occur in other species. However, there are few data in this regard to make an adequate assessment.

3.2
OVLT

Initially the OVLT was divided into two zones, an external pericisternal zone and an internal periventricular zone (Hofer 1958; Kuhlenbeck 1970; Weindl et al. 1967). Us-

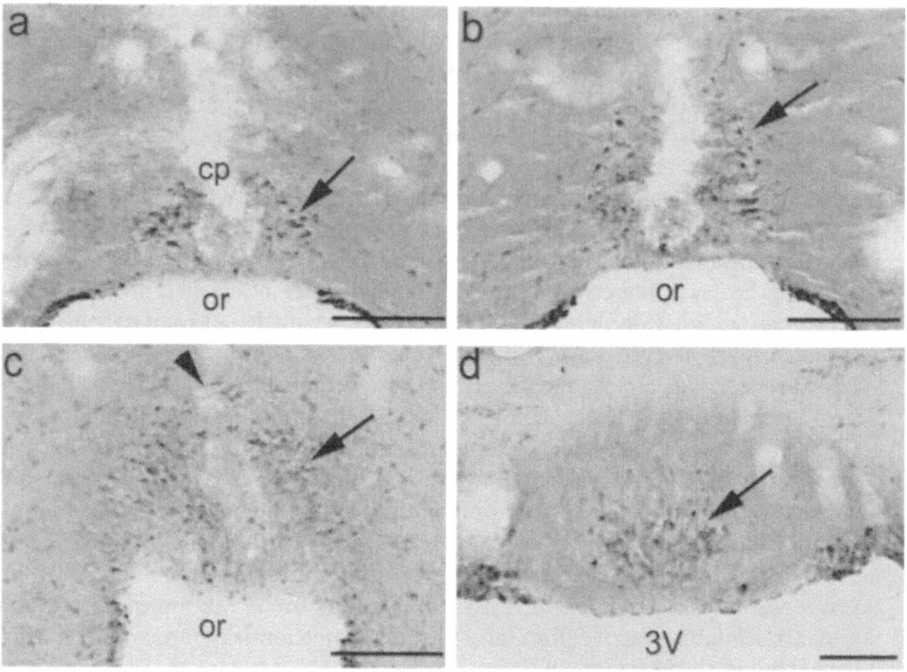

Fig. 12a–d Immunohistochemical labelling of calbindin-containing neurons. **a–c** Calbindin immunoreactivity in the lateral zone of the OVLT (*arrow*), which appears to be continuous with the most caudal periventricular part of the OVLT that can be seen in **c**. The dorsal cap *(arrowhead)* is not calbindin positive. The most rostral section is shown in **a** and the most caudal in **c**. **d** Calbindin immunoreactivity in the ventromedial core (*arrow*) but not the outer shell of the subfornical organ. *cp*, Capillary plexus; *or*, optic recess of the third ventricle; *OVLT*, organum vasculosum of the lamina terminalis; *3V*, third ventricle. *Calibration bars*=150 μm

ing criteria based on the entry of blood-borne markers into its neuropil, and barriers to the diffusion of markers between different compartments of the OVLT, Krisch et al. (1987) subdivided the OVLT into two different compartments. Although both compartments exhibited capillaries with fenestrated capillary endothelium, circulating horseradish peroxidase did not enter the interstitium of one of these compartments, the region caudal and adjacent to the dense intrapial capillary plexus. The reason for this is not known. Between this region and the third ventricle was a more caudal second compartment into which blood-borne horseradish peroxidase was able to penetrate. On the basis of vascular architecture, neural connectivity, chemical phenotype, receptor density and immediate-early gene expression in response to different stimuli, we propose three functional subdivisions of the rat OVLT (see schematic diagram, Fig. 11). The first is the rostromedial vascular region and immediately adjacent tissue characterised by an intrapial capillary plexus with extensive perivascular spaces and neurosecretory terminals. The second is its arching dorsal cap, which is part of the inner zone and covers the vascularised region of the OVLT along its rostrocaudal extent. Neural projections of this dorsal cap are predominantly to the supraoptic and magnocellular parts of the paraventricular nuclei (see Sect. 6).

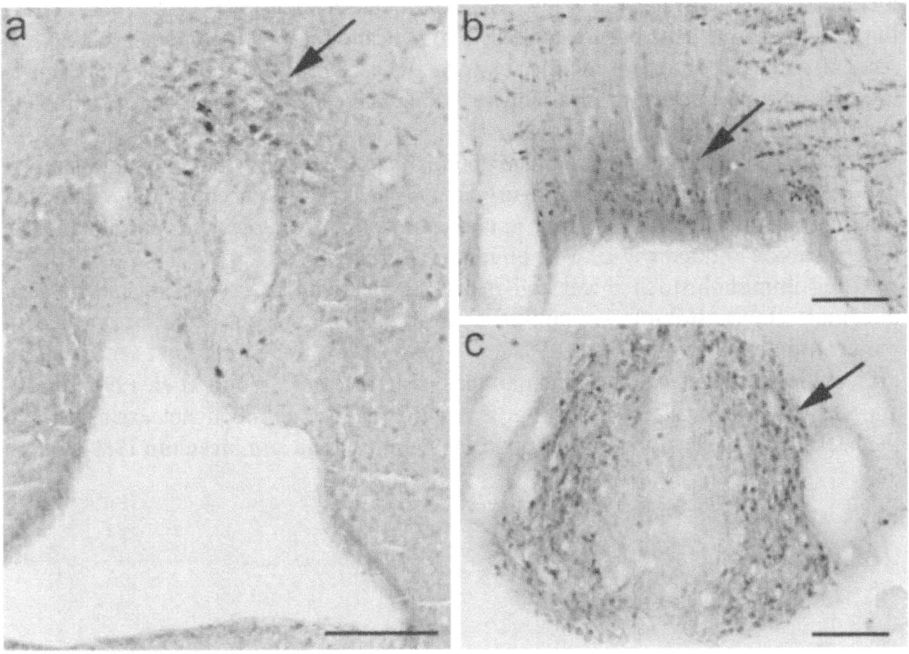

Fig. 13. **a** Immunohistochemical labelling of calretinin-containing neurons in the dorsal cap of the OVLT (*arrow*). **b** Calretinin immunoreactivity in the rostral part of the subfornical organ (*arrow*). **c** Calretinin immunoreactivity in the outer shell of the subfornical organ (*arrow*). Note absence of calretinin immunoreactivity in the lateral zone of the OVLT and the ventromedial core of the subfornical organ. *Calibration bars*=150 μm

Neurons in the dorsal cap are strongly activated by hypertonicity and by the hormone relaxin (McKinley et al. 1998) and are immunoreactive for calretinin (Fig. 13a). The third subdivision of the OVLT is the lateral and posterior periventricular tissue which surrounds the central vascular plexus. This region is rich in angiotensin II AT_1 receptors (Giles et al. 1999; Allen et al. 2000), and neurons in this region are particularly responsive to angiotensin II (McKinley et al. 1998). Efferent neural connections to the lateral hypothalamus, bed nucleus of the stria terminalis and parvocellular divisions of the paraventricular nucleus also characterise this part of the OVLT (see Sect. 6). In addition, many neurons in the lateral and caudal parts of the OVLT exhibit calbindin immunoreactivity (Fig. 12a–c).

3.3 Area Postrema

The area postrema of the rat has been subdivided into three or four zones depending on the investigator and the species. Generally, three morphologically distinct regions have been recognised in dogs, monkeys and cats. These are the dorsal periventricular mantle zone, a central zone and a ventral junctional zone adjacent to the NTS (Chernicky et al. 1980; Brizzee and Klara 1984; Klara and Brizzee 1975; Leslie 1986). These three subdivisions have been extended to four in the rat, in which a lateral zone was recognised in the caudal to middle rostrocaudal aspect of the area postrema. This is in addition to the dorsal, central and ventral zones which extend along its complete length (Gross 1991), and these are shown in Fig. 11.

The dorsal periventricular mantle contains flattened ependyma and supraependymal cells in addition to layers of neurons and glial cells subjacent to the ependyma. The central zone is rich in neurons and axon terminals, whereas the ventral zone has fewer neurons and many glia at its boundary with the NTS.

Using immunohistochemical and neural connectivity as criteria, it can be seen that most cholecystokinin-containing neurons in the area postrema are located in the central zone (Krukoff et al. 1992), as are most of the neurons that send efferent projections to the lateral parabrachial nucleus in the rat (Herbert et al. 1991). It also appears that the vagal afferent input is predominantly, although not exclusively, to the lateral part of the dorsal periventricular mantle (Kalia and Mesulum 1980).

4 Vasculature, Compartmental Barriers, Neurons and Glia in the Sensory Circumventricular Organs

4.1 Subfornical Organ

4.1.1 Arterial Sources and Venous Drainage

The arterial supply and microvasculature of the subfornical organ have been studied in several mammalian species (human, rat, cat, rabbit). Arteries supplying the subfornical enter it both in the ventral stalk as a branch of the preoptic artery and dorsally and caudally from branches of arteries supplying the choroid plexus (Duvernoy and Koritke 1964). In the rat, the subfornical organ receives a major supply from the subfornical artery which branches from the anterior cerebral artery. To reach the subfornical organ, the subfornical artery passes caudally over the splenium of the corpus collosum, returning rostrally under this fibre tract back towards the subfornical organ through the arachnoid tissue (Spoerri 1963). Before entering the dorsal-caudal aspect of the subfornical organ, this artery is joined by branches of the anterior and posterior choroidal arteries. Veins issuing from the choroid plexus also pass through the subfornical organ of the rat (Fig. 14b, d) and probably other species as well (Spoerri 1963; Duvernoy and Koritke 1964). Within the body of the subfornical organ, arterioles both from dorsal and ventral stalks reticulate into an extensive network of thin capillaries which are sometimes looped and reach near to the ventricular surface (Fig. 14b, c). Capillaries drain rostrally and laterally into large medial septal veins which flow dorsally and may reach the great cerebral vein of Galen.

4.1.2 Capillary Plexus

The capillary plexus of rat, cat, rabbit and human subfornical organ has been described in detail (Spoerri 1965; Duvernoy and Koritke 1964, 1969; Shaver et al. 1990; Dellman 1987). In each case a dense, intricate network of capillaries was observed, with sub-ependymal capillary loops present to varying degrees, depending on the species (Fig. 14b, c). The capillaries tend to be of finer diameter in the ventral region of the subfornical organ, and this is particularly evident in the cat (Duvernoy and Koritke 1964). In the rat, the density of capillaries is greater in the ventromedial core (Fig. 14b) than in the outer shell (Bouchaud et al. 1989; Dellman 1998). Capillaries

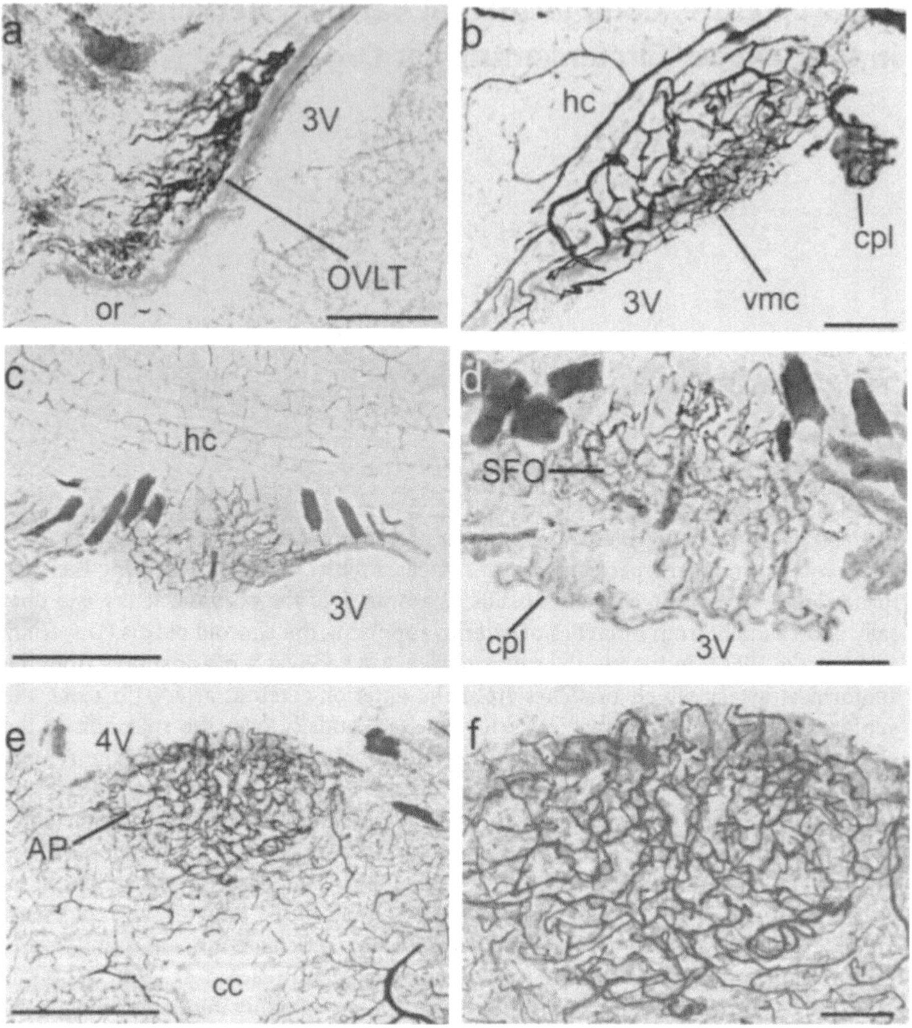

Fig. 14a–f Vasculature of the rat sensory circumventricular organs which has been filled with a suspension of India ink in gelatine after its post-mortem injection into the systemic vasculature. **a** The capillary plexus of the OVLT in mid-sagittal section. **b** Mid-sagittal section showing the extensive vasculature of the subfornical organ. **c**, **d** Coronal sections through the subfornical organ. The most caudal section is shown in **d**, and the vascular connections to the attached choroid plexus can be seen. **e**, **f** Capillaries in the area postrema at two different magnifications. *AP*, area postrema; *cc*, central canal; *cpl*, choroid plexus; *hc*, hippocampal commissure; *or*, optic recess; *SFO*, subfornical organ; *vmc*, ventromedial core of the subfornical organ; *3V*, third ventricle; *4V*, fourth ventricle. *Calibration bars*=300 μm in **a**, 150 μm in **b**, **d**, and **f** and 600 μm in **c** and **e**.

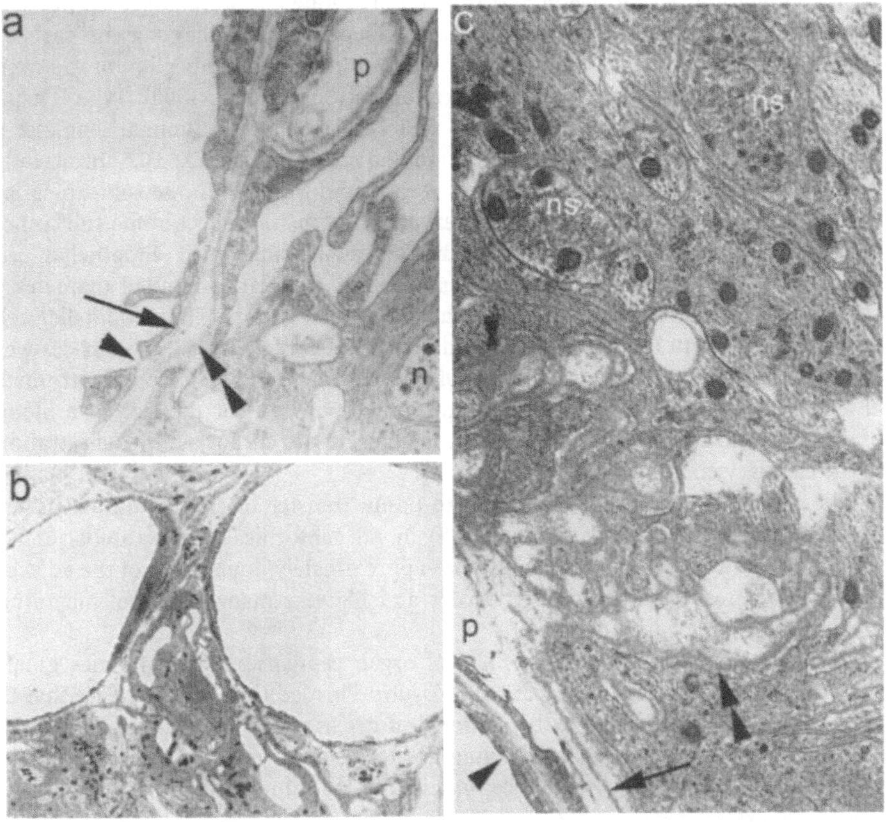

Fig. 15a–c Electron micrographs of fenestrated capillary endothelium in sensory CVOs. **a** Part of the wall of a fenestrated capillary in the subfornical organ of a sheep. An *arrowhead* indicates one of the four fenestrations in this section, an *arrow* indicates the inner basement membrane and a *double arrowhead* indicates the outer basement membrane. The perivascular space (*double arrowheads*) and nerve processes (*n*) are indicated. ×20,000. **b** Two fenestrated blood vessels and extensive perivascular space in the OVLT of a sheep. ×2,000. **c** Neurosecretory nerve terminals (*ns*) surrounded by glial processes in the OVLT of a sheep. Extensive labyrinthine perivascular spaces are enclosed by two basement membranes (*arrow* and *double arrowhead*). A *single arrowhead* indicates a fenestration in the capillary wall. ×20,000. *n*, Nerve process; *ns*, neurosecretory terminals; *p*, perivascular space

may have either fenestrated or non-fenestrated endothelial cells. An example of fenestrations and expanded perivascular space in the subfornical organ (of a sheep) is shown in Fig. 15a.

Shaver et al. (1990) made quantitative analysis of the fine structure of capillaries in rat (Long-Evans strain) subfornical organ and distinguished three particular classes of capillary, types I, II and III. Type I capillaries were described as having a continuous endothelial sheath, many intracellular vesicles ($10\pm1\ \mu m^{-1}$) a lumen diameter of $4.9\pm0.2\ \mu m$, an endothelial wall thickness of $0.3\pm0.02\ \mu m$ and pericapillary space of $28\pm6\ \mu m^2$. Type I capillaries were found predominantly in the transitional zone between the central region and rostral outer zone and were not fenestrated. In regard to the intracellular vesicles, it has been suggested that they may be part of a system of transendothelial channels which allow direct communication of the capillary lumen and neuropil of the subfornical organ (Shaver et al. 1990). Type II capillaries were not fenestrated and had no pericapillary space, a luminal diameter of 5.0±0.3 μm, an endothelial wall thickness of 0.31±0.02 μm and 3.6±0.5 intracellular vesicles and pits/μm^2. Type II capillaries were found throughout the subfornical organ but were the only capillary type observed in the rostral aspect of the subfornical organ. Type III capillaries exhibited discontinuous fenestrated endothelial cells (9±1 μm^{-2}) that were thin walled (0.17±0.01 μm) and had a luminal diameter of 4.4±02 μm and an extensive perivascular space (56±9 μm^2). The endothelial cells contained many intracellular vesicles and pits (13±1 μm^{-2}). Type III capillaries were prevalent in the central and caudal part of the subfornical organ, i.e. the ventromedial core according to our designation, and are responsible for the lack of a blood-brain barrier in this subdivision of the subfornical organ because of the fenestrations in the capillary endothelium.

The blood flow, transit time, blood-to-tissue transfer of circulating agents and permeability × surface area product (PS) in rat subfornical organ capillaries was measured by Gross (1991). Blood flow was approximately double that of the adjacent neuropil, whereas blood-to-tissue transfer and PS were many orders of magnitude greater in the subfornical organ.

Plasma transit time for rat subfornical organ capillaries was 3–4 times greater than for the adjacent hippocampal commissure. This led Gross to speculate that the slow velocity of blood flow in the subfornical organ would be advantageous to the functioning of a sensory organ, allowing greater time for molecules in fluid derived from blood to be in contact with receptors within the subfornical organ.

4.1.3 Compartmental Barriers

As mentioned above, fenestrated capillary endothelium appears only to be present in the caudal core region of the subfornical organ of the rat. Consistent with this observation are data of Krisch et al. (1978), who reported that only part of the subfornical organ of the rat is exposed to the haemal milieu, that being the ventromedial core region. They proposed that tanycyte processes extend from the ependyma to encircle capillaries within the subfornical organ, rather than capillaries at its border with adjacent regions, as is the case with other CVOs. Krisch et al. (1978) showed that adjacent tanycyte processes are connected by helical tight junctions which interrupt the intercellular space between them. They suggest that these tight junctions maintain a barrier between the haemal milieu of the ventromedial core of the subfornical organ and its outer shell. However, whether this barrier effectively excludes circulating molecules from the remainder of the subfornical organ is doubtful. Neurons were labelled with Evans blue throughout the subfornical organ 1 h after its injection into the tail vein of rats (Schmid 1995). We found that 3 min after injection of horseradish peroxidase intravenously, this large protein had diffused throughout the subfornical organ (Fig. 16a). Moreover, experiments in which angiotensin II was infused intra-

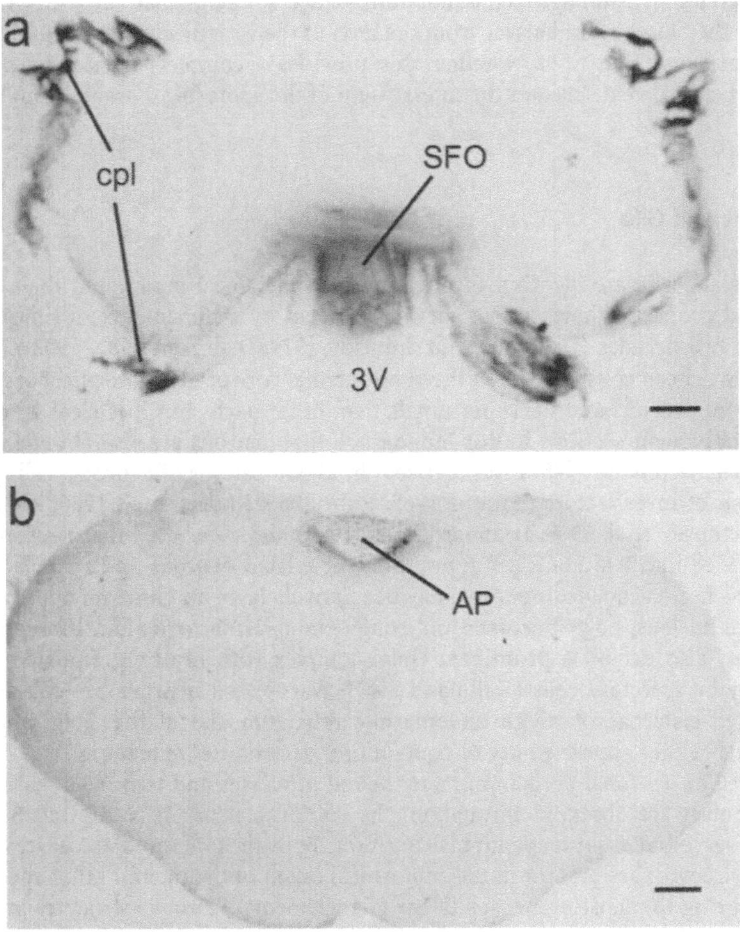

Fig. 16a, b Horseradish peroxidase (HRP) in the subfornical organ, choroid plexus and area postrema, but not other regions of the rat brain, 5 min after an intravenous injection of HRP. **a** Subfornical organ and choroid plexus of the lateral ventricle. **b** Medulla oblongata showing HRP reaction product in the area postrema, especially at its ventral boundary. *AP*, area postrema; *cpl*, choroid plexus; *SFO*, subfornical organ; *3V*, third ventricle. *Calibration bars*=300 μm

venously into conscious rats showed that neurons in both the outer shell of the subfornical organ as well as its ventromedial core were stimulated, as indicated by Fos immunoreactivity (McKinley et al. 1992a). Even more striking is the Fos immunoreactivity that is expressed in the outer shell of the subfornical organ during intravenous infusion of the hormone relaxin (Fig. 9). This Fos expression is almost certainly due to direct stimulation of these neurons by circulating relaxin. These data suggest that peptides with molecular masses varying from 1,000 to 40,000 Da enter into the interstitium of the whole subfornical organ.

Adjacent ependymal cells of the subfornical organ have been reported to have tight junctions connecting them, and also in some instances zonulae adhaerentes

(Dellman 1987). Although tight junctions between ependymal cells in CVOs will transfer the blood-brain barrier from the level of their capillary endothelium to these ependymal tight junctions, whether this provides a complete barrier between the CSF of the third ventricle and the interstitium of the subfornical organ is not clear.

4.1.4 Neurons and Glia

Neurons and glia are plentiful throughout the subfornical organ, and their appearance and properties have been reviewed in detail by Dellman several times during the past two decades (Dellmann and Simpson 1979; Dellmann 1987, 1998). In summary, it has been considered that the ventromedial core of the subfornical organ contains more neurons and nerve terminals than other parts, but it is clear from many of the subsequent sections in this monograph that neurons are also abundant in the outer shell. Dellmann (1987) summarised the classifications of neuronal cell types by a number of investigators (Andres 1965; Rohr 1966; Rudert et al. 1966; Akert et al. 1967; Dempsey et al. 1968; Leonhardt et al. 1969) and concluded that neurons could be classified into two main cell types: non-vacuolated neurons and vacuolated neurons. The non-vacuolated neurons may be relatively large and have an approximately spherical nucleus, polyribosomes and rough endoplasmic reticulum throughout the cell. They also exhibit a prominent Golgi complex with nearby granulated vesicles and may be secretory cells (Dellmann 1987). Vacuolated neurons are characterised by dilated cisternae of rough endoplasmic reticulum and of the Golgi apparatus, which may either appear empty or contain fine electron-dense material.

As well as neuronal perikarya, many neural processes and terminals making synaptic contact are observed throughout the subfornical organ of the rat. Some are even observed at a supraependymal location. Both protoplasmic astrocytes and fibrous astrocytes are present in the subfornical organ of the rat and other species, the former being the most numerous of the glial elements. Fibrous astrocytes are sometimes termed 'satellite cells' because of their close association with neurons, axons and dendrites (Dellmann 1987).

4.2 OVLT

4.2.1 Arterial Sources

The vascular supply and intrinsic vasculature of the OVLT, as with other CVOs, have been studied in many species with light and electron microscopy and after intravascular injections of ink, dyes and vascular casts. In all mammals studied, the OVLT exhibits a remarkably rich plexus of blood vessels (Fig. 14a), and a general pattern of arrangement of these vessels emerges which is shown diagrammatically in Fig. 17. The arterial supply of the OVLT of human, macaque, rat and fox arises rostrally from one or two small preoptic arteries that branch off usually from the anterior commu-

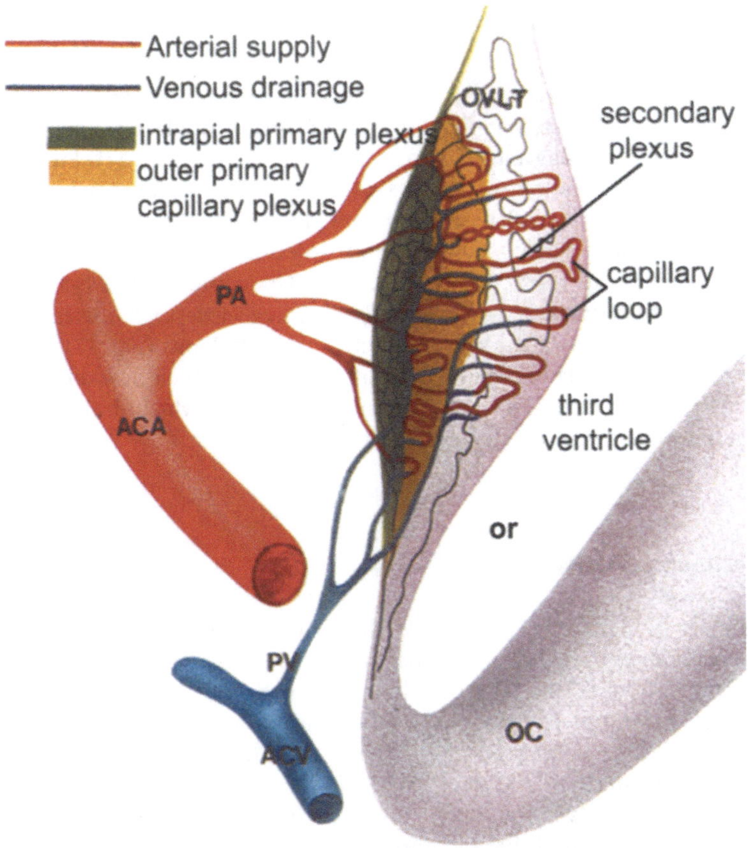

Fig. 17 Schematic representation of the general vascular arrangements in the mammalian OVLT. The preoptic arteries supplying the OVLT branch from the anterior communicating artery. A primary capillary plexus arises from the intrapial vessels, and secondary capillaries extend towards the ventricular surface, often looping and twisting. The venous drainage is directed caudally and laterally into preoptic veins. *ACA*, anterior communicating artery; *or*, optic recess of the third ventricle; *OVLT*, organum vasculosum of the lamina terminalis; *PA*, preoptic arteries; *PV*, preoptic vein

nicating artery and/or the anterior cerebral artery (Mergner 1959; Duvernoy and Koritke 1964). In the human brain, Duvernoy et al. (1969) describe four arterial supplies to the OVLT: a superior median source which branches from the anterior communicating artery; two preoptic arteries issuing from each anterior cerebral artery that divide into branches to supply the vascular plexus of the OVLT from its rostral aspect and an inferior arterial source to the OVLT coming medially from below the optic chiasm.

4.2.2
Capillary Plexus and Venous Drainage

The arterial blood supply to the OVLT enters the rostral pial covering of the OVLT and ramifies into a dense superficial intrapial primary capillary plexus and an outer primary plexus extending into the body of this CVO (Figs. 2b, 17). From these primary vascular plexuses there arises a secondary capillary plexus, whose capillaries penetrate the body of the OVLT as convoluted loops which are more pronounced in some species than in others such as the rat (Mergner 1959; Duvernoy and Koritke 1964; Duvernoy et al. 1969; Weindl et al. 1967; Grafe and Weindl 1987). Some fine vessels leave the primary plexus caudally, reaching close to the ependyma of the third ventricle—the so-called sub-ependymal capillary loops (Fig. 17). The OVLT of the rat is drained caudally by four to eight small veins which flow dorsally to a 'venous tetrapod' and ventrally to the prechiasmatic sinus, which drains eventually to the superior sagittal sinus (Grafe and Weindl 1987).

4.2.3
Compartmental Barriers

Early studies by Behnsen (1927) and Wislocki and Leduc (1952) showed that, unlike most brain tissue, intravitally administered trypan blue enters the OVLT from the bloodstream, showing the lack of a blood-brain barrier in this CVO. Subsequent studies, using the protein horseradish peroxidase as a marker of passage across the blood–brain barrier, confirmed its entry into the interstitium of the OVLT from the bloodstream (Weindl 1973; Krisch 1978; McKinley et al. 1990). Electron microscopic studies (Krisch et al. 1987) of the various vascular compartments and barriers within the OVLT of the rat found that there were two distinct functional compartments, although fenestrated endothelial cells line capillaries in both regions. The outer compartment was adjacent to and included the intrapial primary capillary plexus. This was separated from an inner periventricular compartment by tanycyte processes joined by helical tight junctions and by processes from meningeal cells. Both compartments were also separated from the CSF of the outer subarachnoid space by a basement membrane. Yamaguchi et al. (2000) compared the permeability of the OVLT of rats and rabbits and concluded that the rabbit OVLT was more accessible to blood-borne molecules than the same region in the rat.

4.2.4
Neurons and Glia

The neurons and glia as well as the various membranes and processes which make up the neuropil of the OVLT have been studied in rabbit (Wiendl et al. 1968, Yamagachi et al. 1993), rat (Rohlich and Wenger 1969; Schwendemann 1973; Le Beux 1972; Leveque 1967; Krisch et al. 1987) human (Wenger and Toro 1971; Kuhlenbeck 1970; Landas and Phillips 1987) and sheep (McKinley et al. 1983a) and reviewed in detail by Leonhardt (1980). General features of the OVLT that have been described in these various mammals are rich vascularisation of the OVLT, flattened ependymal cells and processes, neuronal perikarya, axons, dendrites, extensive perivascular spaces, tanycyte processes and neurosecretory terminals.

The OVLT of the rabbit has been described in the most detail, and its internal zone contains many neurons. These neurons of diameter 7–10 μm exhibit axo-somatic synapses, and many neurons are vacuolated as are some cell processes. Astrocytes, their processes, and to a lesser extent other glia, are found throughout the OVLT. Astrocytic processes, axons and dendrites form the neuropil, in which many axo-dendritic synapses can be observed. The rostral part of the OVLT is composed largely of astrocytic process which often contain bundles of tightly packed filaments. This also appears to be the case in humans, rats and sheep; however, Wenger and Toro (1971) did not observe neuronal perikarya in this region of the OVLT of post-mortem human tissue. Perivascular 'lakes' often close to the third ventricular surface have been observed in rats and rabbits (Weindl et al. 1967; Schwendemann 1973, Yamaguchi et al. 1993).

The rostral outer surface of the OVLT is covered in basement membrane, and as described above, the tissue it borders invaginates into a labyrinthine system of blood vessels, astrocytic processes, basement membranes and perivascular spaces (Fig. 15b, c), often containing connective tissue elements such as collagen fibrils.

Nerve fibres coming from outside the OVLT, which contain large dense core vesicles or granulated vesicles, are often observed to run in the dorsal–ventral direction between ependymal tanycyte processes and to terminate with such tanycyte processes next to perivascular spaces. They do not make synaptic contacts but have free nerve-endings and have the appearance of neurosecretory terminals (Fig. 15c). It is likely that many of these are LHRH-secreting terminals. Indeed, ablation of the medial preoptic region, which contains many of the parent cell bodies of the LHRH-containing fibres, results in the degeneration of LHRH-containing terminals in the OVLT (Palkovits et al. 1978). Some free neurosecretory terminals have also been observed to terminate at the ventricular surface of the preoptic recess in the rat (Le Beux 1972).

In addition to free nerve endings, many axo-dendritic synaptic profiles are seen within the neuropil of the OVLT, and these exhibit smaller electron-lucent vesicles near synaptic specialisations (Le Beux 1972). The elongated basal process of ependymal tanycytes often extend in groups into the body of the OVLT to reach perivascular spaces in or adjacent to the OVLT. Krisch et al. (1987) described tight junctions joining these processes, which they considered to make up a barrier between the OVLT and adjacent brain regions, thus preserving the blood-brain barrier of those adjacent regions.

The intrapial primary plexus of the rabbit often exhibits extended perivascular spaces surrounded by basement membrane, through which glial and neural processes containing secretory vesicles enter the vascular zone (Weindl et al. 1967). The secondary capillary plexus of the inner zone is less developed in the rat than in the rabbit (Yamaguchi et al. 1993).

In the rat, the perivascular space surrounding the fenestrated capillary endothelium in the inner zone of the OVLT is surrounded by two basement membranes an inner basement membrane around the endothelium and an outer basement membrane at which tanycyte and neuro-secretory terminals may abut. A perivascular space containing adventitial cells and connective tissue separates these inner and outer basement membranes (Schwendemann 1973).

In the rabbit, Weindl et al. (1967) distinguished four types of capillary loops which extend into the inner zone of the OVLT. This classification was based on the size of their perivascular space, which may range from quite narrow to an extremely wide perivascular lake located immediately beneath the ependyma.

4.3 Area Postrema

4.3.1 Arterial Sources and Venous Drainage

The main arterial source for the area postrema, which has been well described in a number of species (rat, cat, human), is the anterior inferior cerebellar artery. It courses along the pial surface of the medulla, lateral and dorsal to the area postrema, and branches twig off and enter the lateral boundary of this CVO adjacent to its attachment to the tela choroidea (Dempsey 1973; Duvernoy and Koritke 1964). In the rat, the area postrema's blood supply comes from an arterial semicircle surrounding its caudal end. As well as branches from the anterior inferior cerebellar artery, this semicircle receives arterial branches from the dorsal rami of the vertebral artery and from an intraspinal network (Kroidl 1968). In the cat, arterial sources also come from the posterior and anterior spinal arteries which anastomose with the posterior inferior cerebellar artery (Brizzee and Neal, 1954). Having entered the area postrema, these arteries arborise into a dense capillary bed. In the rat some capillaries connect to the choroid plexus of the fourth ventricle (Kroidl 1968). The main drainage from the capillary bed of the area postrema is to veins deep within the area postrema and adjacent to the underlying medullary neuropil. In the cat, these veins issue into two larger veins, one at the dorsal and one at the ventral pole of the area postrema.

4.3.2 Capillary Plexus

Similar to the subfornical organ and OVLT, the capillary network of the area postrema is extensive and tortuous (Fig. 14e, f), with ribbon-like capillaries branching, interdigitating, looping and entwining (Roth and Yamamoto 1968). In the rat, the density and tortuosity of these capillaries was reported to increase from rostral to caudal (Roth and Yamamoto 1968). However, quantitative analysis by Shaver et al. (1990) showed that in general, capillary density across the area postrema was reasonably uniform at approximately $500/mm^2$, except in the most ventral part where it fell to $300/mm^2$. Dempsey (1973) reported that thin-walled capillaries with relatively wide lumens (described as sinusoidal in nature) connected to finer capillaries with fenestrated endothelia containing many mitochondria and pinocytotic vesicles. Quantitative analysis of the capillaries of the area postrema of the rat showed two categories of capillaries—type I with diameters of 3–7 μm, and type II of diameter

7–15 μm. Both types of capillary have wide perivascular spaces, bounded on the inner side by endothelial cells and on the outer side by basement membranes of pial origin. In some species, subependymal capillary loops are evident. Gross (1991) made calculations of plasma flow, blood-to-tissue transfer and extraction fraction of a neutral amino acid, and capillary permeability in rat area postrema. He found that plasma flow was 50% greater than in the adjacent medullary regions and that unidirectional blood-to-tissue transfer, and extraction fraction and permeability × surface area (PS) product were more than 150 times those of adjacent regions. Another interesting observation derived from these measurements was that the calculated velocity of blood flow in the capillaries of the area postrema is relatively slow, with the transit time for capillary blood flow being more than double that of the dorsal motor nucleus of the vagus. Gross (1991) concluded that this slow transit time of plasma in the capillaries of the area postrema (and also subfornical organ) would permit a greater time for it to be in contact with fenestrations and therefore obtain access to the receptors and sensors within the area postrema.

4.3.3
Compartmental Barriers

Several early workers observed that intravital dyes stained the area postrema but not the surrounding medullary regions (Wislocki and King 1936; Wislocki and Leduc 1952). Later, Krisch (1986) observed that the interstitium of the area postrema was marked with HRP 5 min after its systemic administration, showing that the blood-brain barrier was absent in the area postrema because of the fenestrated endothelium forming the walls of many of its capillaries. In regard to the separation of the area postrema from adjacent neuropil, Krisch et al. (1978) examined the non-fenestrated capillaries of the subpostremal area using conventional and freeze-etching electron microscopy. They observed that tanycytes attached by tight junctions surrounded these capillaries. They have proposed that the haemal milieu of the interstitum of the area postrema is separated from the subpostremal region and NTS by rows of tanycyte processes joined by helically arranged tight junctions. Krisch (1986) investigated the border region between the haemal milieu of the area postrema and the outer region of the CSF milieu of the leptomeningeal compartment and concluded that leptomeningeal cells of the pia mater, connected by tight junctions and probably a flocculent ground substance, establish a barrier between the area postrema and the meningeal compartment.

4.3.4
Neurons and Glia

Within the neuropil of the area postrema, neurons and glial cells are found together with many glial and non-myelinated nerve processes and nerve endings. Neural elements are usually separated from each other by surrounding glial processes (Dempsey 1973), but axo-somatic and axo-dendritic synapses are common. Dempsey (1973) observed that neurons, nerve terminals and fibres are also within the perivascular sheaths of capillaries within the area postrema of the rat and suggested that these cells subserved neuroendocrine functions. These neural elements are not surrounded by glial processes and are thus directly exposed to the perivascular fluid. Collagenous fibrils, macrophages and fibroblasts are also observed within the perivascular sheaths. Basement membranes surround the perivascular sheaths (Rohrschneider et al. 1972).

5 Neurochemical Aspects of Sensory Circumventricular Organs

5.1 Receptors

For the sensory CVOs to be responsive to circulating factors such as various hormones, peptides or ions, not only do they need the blood-brain barrier to be absent so that they are exposed to this environment, but specific receptors which bind these agents are necessary on the cell surface to specifically detect them.

Autoradiographic binding studies, immunohistochemistry, in situ hybridisation and neurophysiological and pharmacological studies reveal that receptors for a number of circulating hormones, neurotransmitters and neuropeptides are present in the sensory CVOs. These are summarised in Table 2 and show that some ligands may have receptors in all three sensory CVOs. Whether or not these receptors are associated with neuronal cell bodies, neural terminals or glial cells is not always evident, particularly if autoradiographic binding techniques have been used to identify receptor binding sites.

5.1.1 Acetylcholine

Evidence of cholinergic receptors in the CVOs comes initially from electrophysiological and pharmacological experiments. Microiontophoretic application of acetylcholine directly onto neurons in the subfornical organ stimulates electrical activity which is mediated by muscarinic receptors (Felix 1976). Stimulation of cholinergic receptors in the subfornical organ by direct injection of carbachol into this CVO initiates drinking and an increased blood pressure which can be blocked by the muscarinic receptor antagonist atropine (Mangiapane and Simpson 1983). Recent studies utilising immunohistochemistry and calcium imaging of isolated subfornical organ neurons indicate that the muscarinic M1 subtype mediates cholinergic actions in the subfornical organ. Presynaptic muscarinic receptors may inhibit GABA release, whereas post-synaptic muscarinic receptors open non-selective cation channels (Johnson et al. 2001). There have been few studies of cholinergic receptors in the OVLT and area postrema; however, acetylcholine has been shown to directly stimulate neurons in the area postrema of the dog (Carpenter et al. 1988) and there is evidence of presynaptic muscarinic M1 receptors on vagal afferents within the area postrema (Reynolds et al. 1994). An agonist-binding subunit of the nicotinic receptor

Table 2 Summary of receptors or binding sites identified in sensory CVOs[a]

	SFO	OVLT	AP
Acetylcholine	+		+
Amylin	+++	+++	+++
Angiotensin II AT_1	+++	+++	+++
Bradykinin B_2	++	++	++
Calcitonin	+++	+++	+++
CGRP	+++	+++	+++
Calcium	+++		
Cholecystokinin CCK-A and -B			+++
Endothelin ET_A or ET_B	+	+	+
GABA-A	++	++	++
Galanin GAL-R1, GAL-R2	++		
Imidazoline	++	++	++
Neurokinin NK_1			+
GLP-1	+++		+++
Glutamate NMDA or AMPA	+++	+++	+++
Lipopolysaccharide CD14	+++	+++	+++
Natriuretic peptide NPR-A, NPR-B	+++	+++	+++
Neuropeptide Y	+++		+
Prostaglandins EP2, EP3, EP4	+++	+++	+++
ATP P2X			+
Relaxin	+++	+++	
Serotonin	++		+
Somatostatin	++		
Steroids			
Estrogen	++	++	++
Glucocorticoid	++	++	++
Progesterone		++	
Testosterone	++	++	++
Vasopressin V1	+++		+++

[a] Density of receptors: +++, high density; ++, medium density; +, low density.

has been identified by in situ hybridisation histochemistry in the area postrema (Wada et al. 1990).

5.1.2 Angiotensin

Angiotensin binding sites and receptors have been widely studied in the sensory CVOs. Early physiological studies of circulating angiotensin's action on the brain showed that its central pressor action was mediated via an action on the area postrema in dogs and its dipsogenic effects through stimulation of the subfornical organ (Joy and Lowe, 1970; Simpson and Routtenberg 1973). Although it had been shown that fluorescent labelled angiotensin II administered into the bloodstream could bind to circumventricular organs (van Houtten et al. 1983), the first comprehensive mapping of angiotensin receptors in the CNS utilised in vitro autoradiographic binding techniques in rats and showed that binding sites for a radiolabelled analogue of angiotensin II were present in the subfornical organ, OVLT and area postrema, as well as a number of other regions of the brain (Mendelsohn et al. 1984). Subsequent identification of the AT_1 and AT_2 receptor subtypes led to the recognition that the AT_1 subtype predominated in the CVOs as well as many other regions (Song et al. 1992; Allen et al. 2000).

In situ hybridisation and immunohistochemistry have confirmed this observation (Lenkei et al. 1997; Giles et al. 1999). These studies in the rat also show that the density of AT_1 receptors within the subfornical organ and OVLT is not uniform. The highest density of AT_1 receptor binding, immunoreactivity or mRNA expression is in the ventromedial core of the subfornical organ, with lower levels in its outer shell (Lenkei et al. 1995; McKinley et al. 1998; Giles et al. 1999). In the OVLT, AT_1 receptors are located predominantly around the lateral edges and caudal periventricular region (McKinley et al. 1998, Giles et al. 1999; Allen et al. 2000). Studies of Fos expression show that these regions are stimulated by circulating angiotensin II (see Sect. 7.1). AT_1 receptors are more uniformly distributed in the area postrema (Allen et al. 2000).

Angiotensin receptors also have been shown to be present at relatively high concentrations in the three sensory CVOs of a number of other mammals: dog (Speth et al. 1985), sheep (McKinley et al. 1986a), rabbit (Mendelsohn et al. 1988) and marsupials such as the grey kangaroo or Northern quoll (M.J. McKinley and A.M. Allen, unpublished observations). In the human brain (Fig. 18a–d), angiotensin receptor binding sites are observed in the subfornical organ and OVLT (McKinley et al. 1987a), but in the area postrema they were undetectable (Fig. 18e; Allen et al. 1988b).

Angiotensin receptor binding in the subfornical organ of rats has been shown to be downregulated by sodium depletion (Ray et al. 1990b) and upregulated by chloride depletion or dehydration (Ray et al. 1990a; Barth and Gerstberger 1999). It has been shown recently that acute sodium deficiency produced by furosemide administration increases the turnover of AT_1 receptor mRNA in the rat subfornical organ, but not the OVLT, at 4 h after the sodium loss and increased the AT_1 receptor binding by 12 h post-diuretic treatment (Charron et al. 2002). In spontaneously hypertensive rats (SHR), angiotensin binding is increased compared to normotensives (Saavedra et al. 1986), and if the SHR is treated with captopril for 2 weeks, the density of angiotensin receptor binding in the subfornical organ decreases considerably (Nazarali et al. 1990).

An interesting aspect of the distribution of AT_1 receptors in rat brain is their existence in many (although not all) of the regions which have direct efferent neural connections from the subfornical organ, OVLT and area postrema. In the case of the subfornical organ and OVLT, these include the median preoptic nucleus, parvocellular regions of the hypothalamic paraventricular nucleus (PVN), the bed nucleus of the stria terminalis, suprachiasmatic nucleus and lateral preoptic nucleus (Allen et al. 2000). In the case of projections from the area postrema, these include the nucleus of the solitary tract, ventrolateral medulla and lateral parabrachial nucleus.

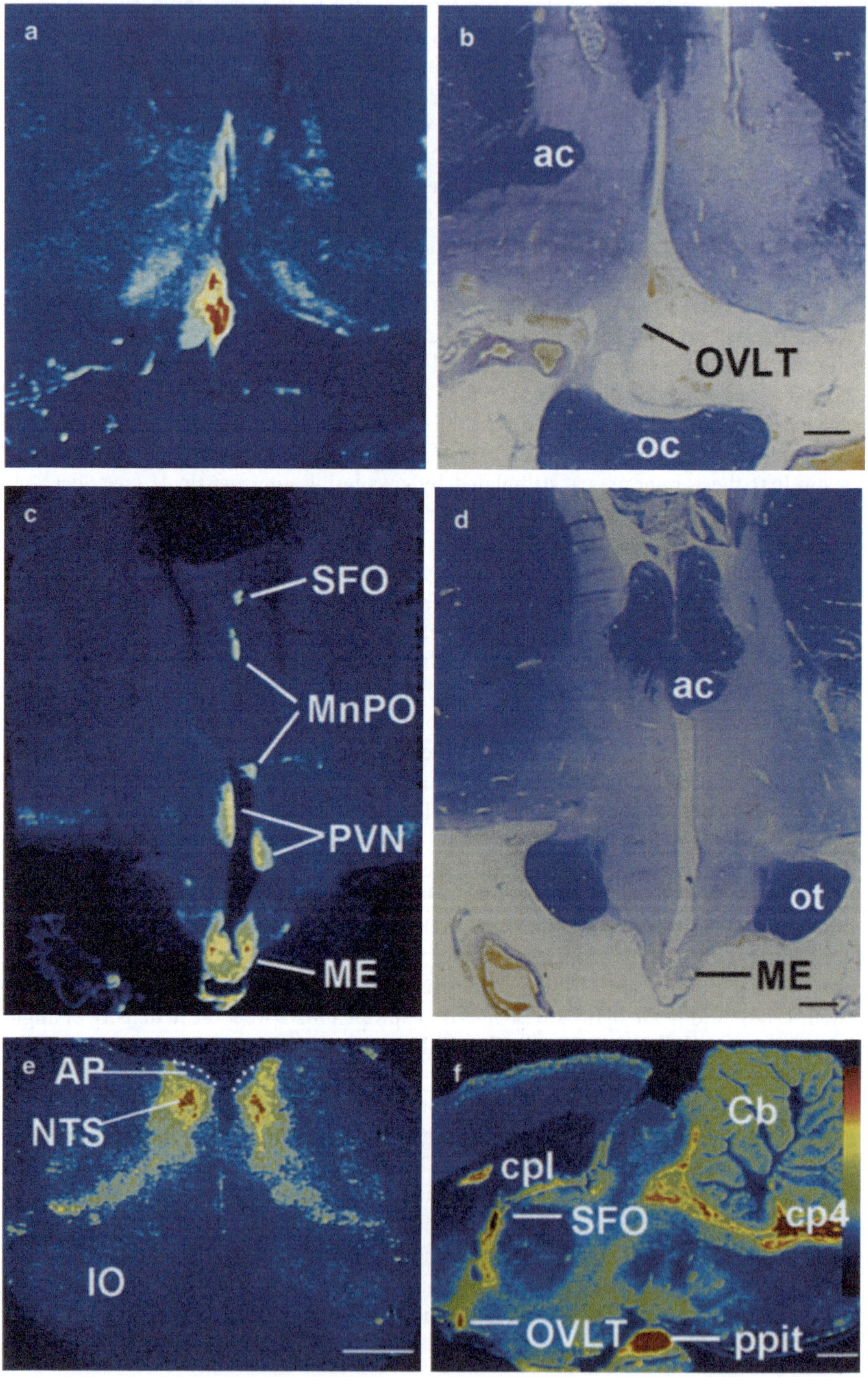
a
b
ac
OVLT
oc
c
SFO
MnPO
PVN
ME
d
ac
ot
ME
e
AP
NTS
IO
f
Cb
cpl
SFO
cp4
OVLT
ppit

5.1.3
Bradykinin

Autoradiographic binding studies using a radiolabelled analogue of the bradykinin B_2 receptor antagonist in sheep and guinea pig have revealed high densities of binding sites for bradykinin in the area postrema and subfornical organ (Fig. 19) and a moderate density in the OVLT (Murone et al. 1997). The cell type(s) (neurons, glia or vasculature) that express the B_2 receptors is yet to be determined.

5.1.4
Calcium

A G protein-coupled seven transmembrane-spanning domain receptor that binds calcium ions has been identified on the membranes of cells in the parathyroid gland, and this is considered to be the calcium receptor regulating the secretion of parathyroid hormone from these glands. Using in situ hybridisation techniques, Rogers et al. (1997) showed that this calcium receptor is also located within the subfornical organ and may be able to act as a sensor of ionic strength (Quinn et al. 1988). Electrophysiological studies indicate that an allosteric agonist molecule for this receptor, NPS R-467, as well as increased $CaCl_2$ concentration, depolarises neurons within the subfornical organ. Microinjection of such calcium receptor agonists into the subfornical organ increases arterial pressure, indicating a possible role in cardiovascular regulation for calcium receptors in the subfornical organ (Washburn et al. 1999b)

5.1.5
Calcitonin, Calcitonin Gene-Related Peptide, Amylin and Adrenomedullin

Calcitonin, calcitonin gene-related peptide (CGRP), amylin and adrenomedullin are a group of related peptides. CGRP is synthesised by alternative processing of the

Fig. 18a–f Angiotensin II receptor binding in the human brain. Computer-generated pseudocolour images of autoradiographs of in vitro binding of [^{125}I][Sar^1-Ile^8] angiotensin II to coronal sections of the OVLT (**a**) and subfornical organ (**c**) but not to the area postrema (**e**) of the human brain. The ventricular boundary of the area postrema is shown by the *white dotted line*. **b**, **d** Adjacent sections stained with luxol fast blue and cresyl violet, showing normal anatomy. **f** Angiotensin-converting enzyme (ACE) localisation in a mid-sagittal section of the rat brain as shown by the binding of [^{125}I]351A, a derivative of the ACE inhibitor lysinopril, to the subfornical organ, OVLT, posterior pituitary and choroid plexus. The choroid plexus of the fourth ventricle obscures the binding to the area postrema. *Colour bar* indicates that the highest density of binding is represented by *red* and the lowest by *blue*. *ac*, anterior commissure; *AP*, area postrema; *Cb*, cerebellum; *cpl*, choroid plexus; *cp4*, choroid plexus of the fourth ventricle; *IO*, inferior olivary nucleus; *ME*, median eminence; *MnPO*, median preoptic nucleus; *NTS*, nucleus of the solitary tract; *oc*, optic chiasm; *ot*, optic tract; *OVLT*, organum vasculosum of the lamina terminalis; *ppit*, posterior pituitary; *PVN*, hypothalamic paraventricular nucleus; *SFO*, subfornical organ. *Calibration bars*=2 mm in **b**, **d**, and **f** and 3 mm in **e**

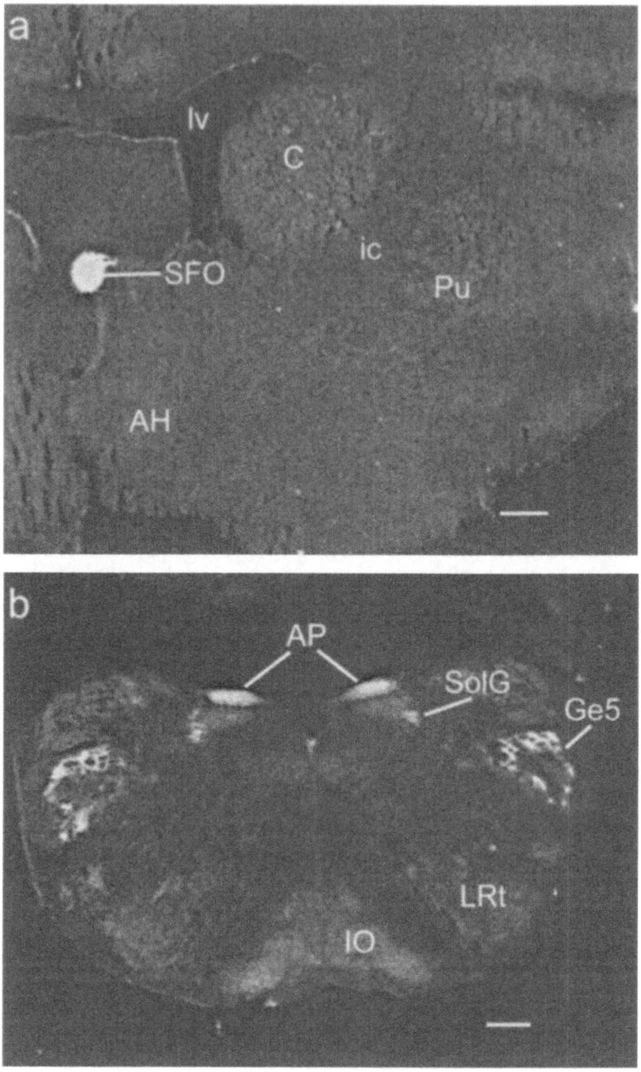

Fig. 19a, b Autoradiographs showing in vitro receptor binding of bradykinin to sensory circumventricular organs in the sheep brain. **a** Coronal sections of the subfornical organ, right side of brain. **b** Coronal section of the area postrema. The ligand used was ^{125}I-Hoe-140, a B_2 receptor antagonist. *AH*, anterior hypothalamus; *AP*, area postrema; *C*, caudate nucleus; *Ge5*, substantia gelatinosus of the spinal trigeminal nucleus; *ic*, internal capsule; *IO*, inferior olivary nucleus; *LRt*, lateral reticular nucleus; *lv*, lateral ventricle; *SFO*, subfornical organ; *SolG*, nucleus of the solitary tract gelatinosus subnucleus. (Reproduced from Murone et al. 1997, with permission). *Calibration bar*=1 mm in **a** and 0.5 mm in **b**

gene encoding calcitonin, and amylin and adrenomedullin have significant sequence homology with CGRP. In addition, it is possible that several of the receptors to which these peptides bind are closely related, because of the different receptor-activated modifying proteins (RAMPs), $RAMP_1$, $RAMP_2$ or $RAMP_3$, that may be co-expressed in the cell membrane with the calcitonin-like receptor (CLR) receptor. RAMPs are able to alter the receptor's affinity for calcitonin, CGRP, amylin or adrenomedullin, depending on which RAMP is co-expressed in the membrane with CLR. It is thus not surprising that each of these ligands may be able to interact with each other's receptors, although with lower affinities than for their own particular receptor (Sexton et al. 2001). The subfornical organ, OVLT and area postrema all exhibit high-affinity in vitro binding for ^{125}I-calcitonin, with K_D values in the picomolar to nanomolar range, as well as comparatively high densities of binding for CGRP and amylin (Henke et al. 1983; Sexton et al. 1986; Sexton et al. 1994a; Sexton et al. 1994b). The functional relevance of calcitonin and CGRP receptors in CVOs remains to be determined.

Electrophysiological data show that both calcitonin and amylin have direct excitatory actions on a high proportion of neurons in the subfornical organ (Riediger et al. 1999). This action is blocked by a selective amylin antagonist, but not by the angiotensin AT_1 antagonist losartan. The observation that amylin induces drinking in rats prompted the suggestion that amylin, which is released from the pancreas during feeding, may play a role in stimulating the drinking associated with feeding (Riediger et al. 1999).

5.1.6 Cholecystokinin

Cholecystokinin (CCK) from the bloodstream has been shown to have a potent anorexic effect. Although this effect may be partially mediated via actions on vagal afferent nerves, there is evidence that it is also partly mediated via CCK receptors in the area postrema.

Binding studies have shown clearly that high-affinity binding sites for labelled CCK occur in the area postrema of the guinea pig but not the other two sensory CVOs (Zarbin et al. 1983). Two main subtypes of CCK receptor have been identified: CCK-A and CCK-B receptors. The CCK-B subtype has been considered to be more prevalent within the CNS and the CCK-A receptors more common in the periphery. There is evidence, however, that both CCK-A and CCK-B receptors are expressed in the area postrema of rats and cynomologous monkeys (Hill and Woodruff 1990; Hill et al. 1990; Mercer et al. 2000).

5.1.7 Endothelin

Two main endothelin receptor subtypes are presently recognised, the ET_A and ET_B receptors. Binding sites for radiolabelled endothelin are present at relatively low density in the subfornical organ and area postrema (Kohzuki et al. 1991), and an immunohistochemical study has identified the ET_A subtype in these two CVOs (Kurakawa et al. 1997). By contrast, the ET_B subtype has been identified in the OVLT by means of immunohistochemistry (Yamamoto et al. 1997). In binding studies, Gebke et al. (2000) observed that astrocytes in the subfornical organ and OVLT expressed both ET_A and ET_B receptor subtypes, and by means of intracellular calcium imaging techniques they showed that these subtypes were functionally active in these astrocytes. Electrophysiological studies in vitro show that endothelin can directly stimulate neurons in the subfornical organ and area postrema, consistent with the presence of ET receptors on neurons in these two sensory CVOs (Ferguson and Bains 1996).

5.1.8
Galanin

Galanin is a broadly distributed neuropeptide in the CNS associated with various monoamine neurotransmitters, including acetylcholine, noradrenaline and serotonin. It usually has inhibitory actions on neurons and has been implicated in nociception and cognition. In vitro autoradiographic studies in rat brain indicate that galanin binding sites are present at moderate levels in the subfornical organ, and possibly also in the OVLT (Skofitsch et al. 1986) and area postrema (Melander et al. 1988). More recent studies indicate that the GALR1 and GALR2 receptor subtypes are both expressed in the subfornical organ of the rat (O'Donnell et al. 1999).

5.1.9
γ-Aminobutyric acid

The amino acid γ-aminobutyric acid (GABA) is a major inhibitory neurotransmitter in sensory CVOs. Immunohistochemical and electrophysiological evidence show GABA-A receptors to be present in the subfornical organ (Weindl et al. 1992; Lee 2001). Receptors for glycine, the other major inhibitory neurotransmitter of the central nervous system, have also been identified in the subfornical organ (Weindl et al. 1992) and area postrema (Jahn et al. 1996).

5.1.10
Glucagon-Like Peptide-1 Amide

Glucagon-like peptide-1 amide (GLP-1), processed from preproglucagon in L-cells in the small intestine, is secreted into the circulation after meals. It is able to lower blood glucose levels because of its stimulation of insulin secretion and suppression of glucagon release. Administered centrally, it has potent inhibitory influences on food and water intake (Tang-Christensen et al. 1996). These latter effects indicate that it may have receptors in the brain. Autoradiographic binding studies utilising ^{125}I-GLP-1 showed that there were binding sites for this peptide at all levels of the neuraxis. Particularly high binding densities for GLP-1 were observed in the area postrema and subfornical organ of the rat. A moderate binding density for this ligand was reported for the anteroventral tip of the third ventricle wall (Göke et al. 1995), which may indicate that the OVLT is also a site of GLP-1 receptors, albeit at lower concentration than the other two sensory CVOs.

In regard to putative GLP-1 receptors in the area postrema, it has been reported that high systemic doses of GLP-1 in human volunteers induce vomiting (cited by Göke et al. 1995). This may be indicative of such receptors in the area postrema, the chemoreceptor trigger zone for vomiting (see Sect. 8.4), being stimulated by GLP-1.

5.1.11 Glutamate

Virtually all neurons in the OVLT (Nelson 1988), subfornical organ (Schmid 1998) and area postrema (Carpenter et al. 1988) are excited by direct iontophoresis or microinjection of glutamate. The excitatory amino acid glutamate acts on both ionotropic and metabotropic receptors in sensory CVOs. In the subfornical organ, electrophysiological data suggest that glutamate is the major excitatory transmitter, its action being mediated by non-NMDA kainate receptors (Xu et al. 2000). Lee et al. (2001) have also shown that presynaptic group II metabotropic glutamate receptors may influence GABAergic inhibitory synapses in the rat subfornical organ. NMDA receptors play an excitatory role in the OVLT (Brann 1995).

In regard to glutamate receptors in the area postrema, evidence exists for the presence of two types of ionotropic receptors. These are the *N*-methyl-D-aspartic acid (NMDA) receptor (Sun and Ferguson 1996) and the α-amino-3-hydroxy-5-methylisoxazole-4-propionic acid (AMPA) receptor (Hay and Lindsley 1999). In the newborn rabbit, ionotropic receptors consist of various combinations of Glu A, Glu B and Glu C subunits (Jahn et al. 1996). Immunohistochemistry indicates that the $GluR_2$ subunit of the AMPA receptor is present in neurons of the area postrema (Lacassagne and Kessler 2000). A metabotropic glutamate receptor, the $mGluR_{2/3}$ receptor, has been identified in the area postrema (Hay et al. 1999).

5.1.12 Imidazoline Receptors

Using a polyclonal antibody to an imidazoline receptor protein, Ruggiero et al. (1998) were able to show immunohistochemically that an imidazoline receptor was present in all three sensory CVOs. At the ultrastructural level, this receptor was localised to presynaptic terminals, glia and neurons in the subfornical organ and OVLT.

5.1.13 Lipopolysaccharide

The membrane CD14 receptor is the receptor to which the bacterial endotoxin lipopolysaccharide, in combination with its binding protein, binds to cells of the immune system such as macrophages and monocytes (Wright et al. 1990). Expression of the CD14 receptor mRNA occurs strongly in subfornical organ, OVLT and area postrema after systemic administration of lipopolysaccharide (Lacroix et al. 1998). Macrophages, microglia and some non-myeloid cells within the CVOs express this receptor. Recent evidence shows that another molecule, Toll-like receptor 4 (TLR-4), may be associated in cell membranes with the CD14 receptor in the sensory CVOs to initiate a signalling cascade (Laflamme and Rivest 2001). These workers showed constitutive expression of TLR-4 mRNA in subfornical organ, OVLT and area postrema, and downregulation of this molecule after administration of lipopolysaccharide (see Sect. 8.5 for discussion of function).

5.1.14
Natriuretic Peptides (ANP, BNP and CNP)

Atrial natriuretic peptide (ANP), brain natriuretic peptide (BNP) and c-type natriuretic peptide (CNP) may all act on receptors in the brain. Circulating ANP and BNP have actions on the brain which include the inhibition of water intake and vasopressin secretion in rats (Antunes-Rodrigues et al. 1985; Samson et al. 1987; Steardo et al. 1994; Ehrlich and Fitts 1990). ANP appears to inhibit the central actions of circulating angiotensin II, and electro-physiological studies have shown direct inhibitory actions of ANP and excitatory actions of CNP on neurons of the subfornical organ in vitro (Schmid and Simon 1992; Ferguson 1999)

Consistent with these results are autoradiographic binding studies in rat and pig brains showing that high-affinity binding sites for ANP or BNP are found in subfornical organ, OVLT and area postrema (Quirion et al. 1986; Niwa et al. 1988; Saavedra et al. 1992). Three natriuretic peptide receptor subtypes have been identified, NPR-A, NPR-B and the so-called clearance receptor, NPR-C. Recent studies using in situ hybridisation techniques to identify NPR-A and NPR-B receptor mRNA show that both of these receptor subtypes are expressed in the subfornical organ and area postrema, whereas the NPR-B subtype is expressed in the OVLT (Saavedra et al. 2000). There is evidence that natriuretic peptide receptors may be upregulated in dehydrated rats (Saavedra et al. 1987) and downregulated in spontaneously hypertensive rats (Saavedra et al. 1986).

5.1.15
Neurokinin Receptors

The neurokinin receptors NK_1, NK_2 and NK_3, at which the neuropeptides substance P, tachykinin A and tachykinin B, respectively, are natural ligands, do not appear to be expressed in subfornical organ or OVLT (Ribeiro-da-Silva 2000). However, autoradiographic results of ^{3}H-substance P binding suggest that the NK_1 receptor may be expressed in area postrema (Ariumi 2000). Substance P has been observed in nerve terminals in the area postrema (Armstrong et al. 1982b) and has been implicated in neural pathways subserving autonomic and sensory function.

5.1.16
Neurotensin

Neurotensin is a neuropeptide that is associated with dopaminergic neurons in several brain regions including dopamine-containing cell groups in the ventral tegmental area and hypothalamus and dopaminergic terminals in the prefrontal cortex of the rat. Receptors for neurotensin have been detected in the subfornical organ of the rat brain by means of the in situ hybridisation technique. Relatively high concentrations of mRNA encoding the type 2 receptor (NT-2R) were observed in this CVO (Walker et al. 1998).

5.1.17 Neuropeptide Y, Peptide YY, Pancreatic Polypeptide

Neuropeptide Y is a 36-amino acid polypeptide that is found in many neural pathways in the brain and is often associated with noradrenaline. It has been implicated in the regulation of appetite and autonomic pathways. Five classes of receptors mediate the biological effects of neuropeptide Y (NPY), peptide YY (PYY) and pancreatic polypeptide (PP). Bolton-Hunter labelled NPY binds strongly to the subfornical organ while binding at low density to the area postrema of the rat (Quirion and Martel 1992). To determine which of the receptor subtypes were binding NPY, Dumont et al. (2000) used a variety of ligands to identify Y_1, Y_2, Y_4 and Y_5 receptor subtype binding. In the subfornical organ Y_1 and Y_2 subtypes are present, whereas in the area postrema the highest density of binding was obtained with ligands selective for Y_2 and Y_4 binding. Moderate binding to the area postrema was obtained with Y_1- and Y_5-selective ligands (Dumont et al. 2000).

5.1.18 Prostaglandins

Binding sites for prostaglandin E_2 (PGE_2) were observed in or very close to the OVLT in rat brain by using tritiated PGE_2 as ligand in autoradiographic studies (Matsumura 1990). With the identification of the different PGE_2 receptors (EP1, EP2, EP3 and EP4), in situ hybridisation studies have shown expression of mRNA encoding these receptors in subfornical organ (EP2, EP3 and EP4 receptors), OVLT (EP3 receptors) and area postrema (EP2 receptors). These may be regulated during immune challenges (Zhang and Rivest 1999; Ek et al. 2000; Matsumura and Kobayashi 2001).

5.1.19 Purinergic Receptors

Purinergic receptors mediate synaptic transmission by ATP, and seven subgroups of purinergic receptors are recognised. One such subgroup of purinergic receptors is the $P2X_2$ receptor, and immunohistochemistry reveals that this receptor is expressed in the area postrema (Atkinson et al. 2000).

5.1.20 Relaxin

Relaxin, a peptide hormone having some structural homology with insulin, is synthesised by the ovaries during the course of pregnancy. Circulating relaxin can influence a number of aspects of CNS function, particularly vasopressin secretion and thirst (Geddes and Summerlee 1995; Sinnayah et al. 1999). It is likely that it exerts its central actions via receptors in the subfornical organ and OVLT.

In vitro autoradiographic binding studies using radiolabelled relaxin show that high -affinity binding sites in the rat brain occur in both the subfornical organ and OVLT, as well as in the hypothalamic supraoptic and paraventricular nuclei, neocortex, hippocampus, dorsal tegmental nucleus and neocortex (Osheroff and Phillips 1991). Whether these putative relaxin receptors are upregulated during pregnancy has not been determined.

5.1.21 Serotonin

Electrophysiological studies show that serotonin receptors in the subfornical organ may mediate both excitatory and inhibitory effects on neurons in this CVO. Pharmacological evidence indicates that $5HT_{2A}$ or $5HT_{2C}$ receptor subtypes mediate excitatory responses, whereas $5HT_1$ receptors may mediate the inhibitory effects of serotonin in the subfornical organ (Scrogin et al. 1998). Serotonin has direct excitatory effects on neurons in the area postrema of the dog (Carpenter et al. 1983).

5.1.22 Somatostatin

In general, somatostatin (SRIF) peptides, which are distributed in many regions of the brain, have inhibitory influences on neuronal functional via SST_1 or SST_2 classes of receptors. These receptors may respond to either the 14-amino acid peptide $SRIF_{14}$ or the extended 28-amino acid peptide $SRIF_{28}$. Studies have shown that of the sensory CVOs, the subfornical organ exhibits a moderately high density of binding of both $SRIF_{14}$ and $SRIF_{28}$. Immunohistochemistry suggests that the SST_1 receptor is expressed in the OVLT of the rat (Dournaud et al. 2000). The function of these receptors in sensory CVOs is not known. Krisch (1992) has also reported binding of somatostatin to the OVLT of the rat.

5.1.23 Steroid Receptors

Autoradiographic binding studies using tritiated steroids show that a number of steroids were bound in the sensory CVOs at quite high levels (Stumpf et al. 1992). Although binding of tritiated progesterone was observed in the OVLT, all three sensory CVOs had binding sites for tritiated oestradiol, dihydroxytestosterone and dexamethasone (Stumpf et al. 1992). In situ hybridisation histochemistry or immunohistochemistry show that the oestrogen receptor α is expressed in the subfornical organ of primate, sheep and rat brain (Blurton-Jones et al. 1999; Mufson et al. 1999; Scott et al. 2000). Some neurons in the subfornical organ and OVLT that express the oestrogen receptor have efferent connections to the supraoptic nucleus (Voisin et al. 1997). The subfornical organ expresses glucocorticoid receptor mRNA (Meyer et al. 1998).

5.1.24
Vasopressin and Oxytocin Receptors

Vasopressin (AVP) exerts its biological actions via two main receptor subtypes, the V_1 and V_2 receptors. Vasopressin-, but not oxytocin, binding sites have been identified in sensory CVOs of the rat brain. In vitro binding of ^{3}H-AVP in the presence of non-radioactive [Thr^4, Gly^7]oxytocin was used to identify vasopressin binding sites distinct from binding to oxytocin receptors, and a high density of receptors was observed in the area postrema as well as in the periphery of the subfornical organ, but not in the OVLT. However, Phillips et al. (1988) were able to use a highly specific V_1 receptor antagonist to label all three sensory CVOs and show that these binding sites represent V_1 receptor binding. Electrophysiological studies showing both excitatory and inhibitory actions of vasopressin on neurons in the area postrema and subfornical organ are consistent with the presence of V_1 receptors in these sensory CVOs (Anthes et al. 1997).

5.2
Neurotransmitters, Neuropeptides, Hormones and Enzymes

5.2.1
Acetylcholine

Mapping of the enzyme associated with acetylcholine production, choline acetyltransferase(ChAT), did not detect cholinergic neurons in the sensory CVOs (Butcher 1995). However, Lewis and Shute (1967) identified acetylcholinesterase-staining fibres from the septal region terminating in the subfornical organ and OVLT. The acetylcholinesterase activity in the OVLT is observed particularly in the dorsal cap region (Meszaros et al. 1969). In the subfornical organ, this acetylcholinesterase activity is situated particularly around its dorsolateral outer shell (Lewis and Shute 1967; Meszaros et al. 1969; Lind 1987), indicating cholinergic innervation of this sensory CVO. Cholinergic vagal afferents have been detected in the area postrema of the ferret (Reynolds et al. 1994).

5.2.2
Angiotensin II, Angiotensin III and Angiotensinogen

Angiotensin-containing neurons, fibres and terminals were mapped throughout the brain of the rat by Lind et al. (1983) with immunohistochemical techniques. The antibody used by these investigators (and several others) is immunoreactive to both angiotensin II and angiotensin III; therefore, both peptides will be included in these maps. The subfornical organ, OVLT and area postrema were amongst a number of sites in which angiotensin-positive cell bodies and terminals were found in relatively high quantities. Ultrastructural studies of the angiotensin immunoreactivity in the subfornical organ (Pickel et al. 1986), OVLT and area postrema (Oldfield et al. 1989; Guan et al. 2000) showed that it was present in axon terminals and dendrites. Although it was also observed in presynaptic terminals, many of the immunoreactive terminals found in CVOs were located in the perivascular spaces and these terminals were not associated with a synapse; presumably they were neurosecretory terminals. Although relatively high amounts of angiotensin II were observed in the sensory CVOs, this is probably not the case for its precursor, angiotensinogen. Angiotensinogen mRNA is mostly localised to glia (Stornetta et al. 1986), and studies utilising in situ hybridisation show that there is moderate expression in the subfornical organ and the signal is only slightly above background in the OVLT and area postrema (Lynch et al. 1987). The expression of angiotensinogen mRNA in the subfornical organ is increased, particularly in its outer shell in rats deprived of water for 2 days (Barth and Gerstberger 1999).

5.2.3 Angiotensin-Converting Enzyme

Angiotensin-converting enzyme (ACE) is the enzyme that cleaves two amino acids from the decapeptide angiotensin I, resulting in the production of angiotensin II. It also causes bradykinin breakdown and as well may hydrolyse several other peptides which include substance P, neurotensin, CCK, β-endorphin, dynorphins, enkephalins and LHRH (Rogerson et al. 1995). ACE has been mapped throughout the brains of a number of species by means of either immunohistochemical techniques or autoradiographic binding of radiolabelled ACE inhibitors. ACE is observed in many brain regions including all sensory CVOs in rat (Fig. 18f), human, monkey, sheep and rabbit brain (Saavedra and Chevillard 1982; Allen et al. 1988a; Chai et al. 1986, 1987, 1990, 1991; Rogerson et al. 1995). In fact, the concentrations of ACE in sensory CVOs of the rat are second only to the choroid plexus, which has the highest levels in the body (Brownfield et al. 1982). Pickel et al. (1986) made immunohistochemical studies of the ultrastructural location of ACE in the subfornical organ and showed that it was located on microvilli and basal processes of ependymal cells, the luminal surface of vascular endothelium and astrocytic processes abutting perivascular spaces.

The high concentrations of ACE in sensory CVOs should be conducive to the local formation of angiotensin II within CVOs close to the sites of angiotensin AT_1 receptors there. The high levels of ACE in the sensory CVOs are particularly relevant to the concentrations of ACE inhibitors needed to block the enzyme in these organs. It is clear from several studies that much higher concentrations of captopril, enalapril, and probably other ACE inhibitors, are needed to block ACE in CVOs than in other regions of the body (Thunhorst et al. 1989; McKinley et al. 1997b).

5.2.4 Catecholamines

Formaldehyde fluorescence techniques were originally used to show that a population of catecholamine-containing neurons were located in the area postrema (Fuxe and Owman 1965). These are classified as part of the A2 noradrenergic group of the dorsolateral medulla. Immunohistochemical detection of the synthetic enzyme tyrosine hydroxylase confirms this noradrenergic group of neurons (Armstrong et al. 1982a), which may have projections to the rostroventrolateral medulla (Blessing et al. 1987). Noradrenergic terminals have also been identified in the area postrema (Armstrong et al. 1982a). No catecholamine-containing neurons have been detected in the subfornical organ; however, some noradrenergic projections from the nucleus of the solitary tract and ventrolateral medulla have been shown to terminate in the subfornical organ of the rat (Ciriello et al. 1996b; Kawano and Masuko 2001). Fluorescence techniques do not detect catecholamine-containing cell bodies in the OVLT of the rat; however, they do show catecholamine-positive nerve fibres and terminals in the OVLT, which were considered to be mainly dopaminergic with a few noradrenergic terminals (Bouchaud and Bosler 1986; Landas and Phillips 1987).

5.2.5 Cholecystokinin

CCK-containing neurons have been identified in the area postrema of the rat (Newton et al. 1985b), and these may project to the lateral parabrachial nucleus (Herbert and Saper 1990). CCK-containing fibres are also present in the area postrema (Vanderhaegen 1985).

5.2.6 Enkephalin

Enkephalin-containing neurons and fibres are present in the area postrema (Petrusz et al. 1985). Enkephalin-containing terminals in the area postrema may form axodendritic and axo-axonic synapses in the area postrema. They may also contain serotonin (Armstrong et al. 1984).

5.2.7 Glutamate, GABA and glycine

Numerous glutamatergic nerve terminals have been identified by immunohistochemical techniques in the subfornical organ (Weindl et al. 1992) and area postrema (Walberg and Otterson 1992), indicative of glutamate's neurotransmitter role. Terminals exhibiting GABA and glycine have also been observed in these sensory CVOs (Weindl et al. 1992; Walberg and Ottersen 1992). Glutamate-, GABA- and glycine-immunoreactive cell bodies are also numerous in the area postrema (Walberg and Otterson 1992). Glutamic acid decarboxylase (GAD) is the enzyme that is responsible for most GABA production and is a reliable marker of GABA. Immunohistochemical studies show that GAD-immunoreactive cell bodies are present in the subfornical organ and area postrema, and fibres are present in all three sensory CVOs. The density of GABA-containing axon terminals appears to be highest in the area postrema and subfornical organ and lowest in the OVLT (Mugnaini and Oertel 1985). The glutamate transporter GLAST is utilised by cells (probably astrocytes) in the sensory CVOs (Berger and Hediger 2000).

5.2.8
Luteinising Hormone-Releasing Hormone

Luteinising hormone-releasing hormone (LHRH)-containing nerve fibres and terminals identified immunohistochemically are concentrated in great profusion within the OVLT of rat and other species. Neurosecretory terminals are plentiful around perivascular spaces (Barry et al. 1985). The cell bodies that are the origin of these LHRH-containing fibres and terminals are located in the medial preoptic region adjacent to the OVLT.

The subfornical organ also contains a few 'stray' LHRH-containing neurons and LHRH-positive fibres and terminals (Barry et al. 1985), although they are much fewer in number than those in the OVLT (McKinley et al. 1990). Some LHRH has been reported to be present in the area postrema (Palkovits and Brownstein 1985).

5.2.9
Neurotensin

The 12-amino acid peptide neurotensin has been detected by immunohistochemical methods in neuronal cell bodies and terminals in the area postrema of the rat and the cat (Newton and Maley 1985c; Emson et al. 1985). Although neurotensin was not detected in fibres or cells in subfornical organ or OVLT (Emson et al. 1985), a neurotensin-containing pathway from the arcuate nucleus to the subfornical organ has been identified (Rosas-Arellano et al. 1996b).

5.2.10
Nitric Oxide Synthase

Immunohistochemical studies have shown that neuronal nitric oxide synthase (nNOS), the enzyme responsible for NO production, is present in many neurons of the subfornical organ and OVLT but not in the area postrema (Ceccatelli et al. 1992; Jurzak et al. 1994; Luckman et al. 1997). Both Jurzak et al. (1994) and Alm et al. (1997) determined that nNOS immunoreactivity in these CVOs corresponded with the histochemical staining of nicotinamide adenine dinucleotide phosphate (NADPH)-diaphorase. NADPH is a co-factor for nNOS, and when used in paraformaldehyde-fixed tissue, NADPH-diaphorase staining labels nNOS-containing neurons and fibres (Jurzak et al. 1994; Alm et al. 1997).

In the subfornical organ, most if not all neurons in its outer shell and ventral stalk exhibited strong immunoreactivity for nNOS (Jurzak et al. 1994; Alm et al. 1997) and NADPH-diaphorase staining (Fig. 20b–d). The immunolabelling was present throughout all parts of this CVO and corresponded completely with NADPH-diaphorase staining at different levels of the subfornical organ.

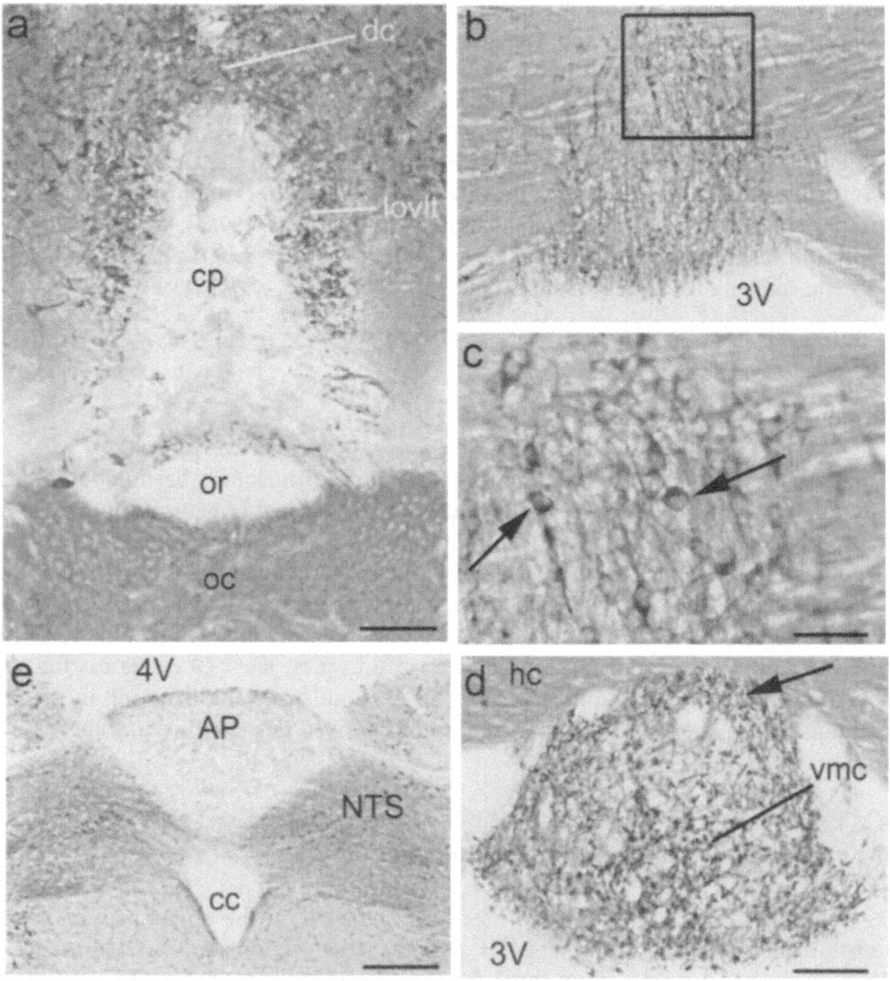

Fig. 20a–e Coronal sections of the sensory circumventricular organs of rat brain showing NADPH-diaphorase staining (which co-localises with and labels neuronal nitric oxide synthase) of neurons and fibres in the sensory circumventricular organs. **a** OVLT. **b** Rostral part of the subfornical organ. **c** High-power magnification of the *square* in **b**, with two NADPH-diaphorase stained neurons indicated by *arrows*. **d** Mid-part of subfornical organ. *Arrow* indicates NADPH-diaphorase-stained neurons in the outer shell. **e** Area postrema; note the lack of labelling of any neurons. *AP*, area postrema; *cc*, central canal; *cp*, capillary plexus; *dc*, dorsal cap; *hc*, hippocampal commissure; *lovlt*, lateral zone of the OVLT; *NTS*, nucleus of the solitary tract; *oc*, optic chiasm; *or*, optic recess of the third ventricle; *OVLT*, organum vasculosum of the lamina terminalis; *vmc*, ventromedial core of subfornical organ; *3V*, third ventricle; *4V*, fourth ventricle. *Calibration bars*=10 μm in **a** and **b**, 30 μm in **c**, 100 μm in **d** and 0.3 mm in **e**

In the OVLT, many nNOS-immunoreactive neurons were located around the rostral primary and secondary capillary plexuses laterally, caudally and dorsally (Alm et al. 1997), and this co-localised exactly with NADPH-diaphorase. Neurons in the OVLT which exhibited NADPH-diaphorase staining often exhibited processes that extended towards the capillary plexus or the ventricular ependyma. This can be seen in Fig. 20a. The NADPH-diaphorase staining is increased in the CVOs of the lamina terminalis in dehydrated rats, indicating that nNOS in the subfornical organ and OVLT may play a role in osmoregulation (Ciriello et al. 1996a). Pharmacological studies in rats indicate that NO may have a function within the OVLT to inhibit sympathetic control of arterial pressure (Lin et al. 1999).

5.2.11 Orexins

The recently identified neuropeptides orexin A and B, which have been implicated in arousal and the regulation of feeding behaviour, are located in neurons in the lateral and posterior hypothalamic regions and perifornical and dorsomedial nuclei of the hypothalamus. These neurons have efferent projections to many regions of the brain, including two of the sensory CVOs, the subfornical organ and area postrema (Nambu et al. 1999). Varicose terminals exhibiting orexin immunoreactivity in these CVOs indicate a likely neurotransmitter role for orexin at these sites. Of interest is the observation that centrally administered orexin A stimulates water drinking in the rat and that prepro-orexin is upregulated in water-deprived rats (Kunii et al. 1999).

5.2.12 Proopiomelanocortin Peptides and Enkephalin

Proopiomelanocortin (POMC), the precursor molecule to the opiate peptide β-endorphin, and also β-lipotropin, α-MSH, adrenocorticotrophin and CLIP, is present in a dense array of neuronal fibres throughout the OVLT, as indicated by immunohistochemical methods. Such fibres were not observed in subfornical organ (Khachaturian et al. 1985).

Enkephalins, opiate peptides that are derived from pre-proenkephalin A, have been detected immunohistochemically in axon terminals in the area postrema of the rat. Some of these terminals also contain serotonin (Pickel 1985).

5.2.13 Serotonin

Immunohistochemical data indicate that serotonin-containing fibres are located in all three sensory CVOs (Fig. 21). These fibres arise from neurons outside the subfornical organ and OVLT and in the case of the subfornical organ originate in the midbrain dorsal and median raphe nuclei (Lind 1986, Bouchaud and Bosler 1986). Although these serotonin-containing terminals are concentrated in the ventromedial

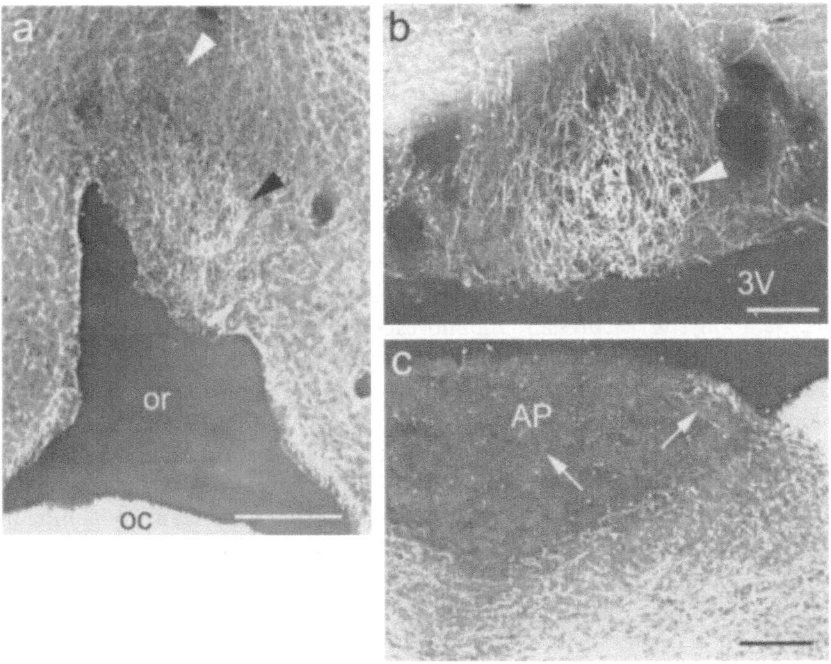

Fig. 21a–c Coronal sections of the sensory circumventricular organs of the rat showing immunohistochemical localisation of the serotonin transporter. **a** OVLT. Extensive immunoreactive fibres and terminals are seen in its lateral zone (*black arrowhead*) but less in the dorsal cap (*white arrowhead*). **b** Subfornical organ. Extensive immunoreactivity is observed in its ventromedial core (*white arrowhead*). **c** Area postrema. Only a few immunoreactive fibres (*arrows*) are detected in the area postrema. *AP*, area postrema; *oc*, optic chiasm; *or*, optic recess of the third ventricle; *OVLT*, organum vasculosum of the lamina terminalis; *3V*, third ventricle. *Calibration bars*=100 μm

core of the subfornical organ, there are serotonergic inputs to its outer shell. This is illustrated in Fig. 21b, which shows immunohistochemically identified serotonin transporter in terminals throughout the subfornical organ but concentrated in the ventromedial core. In the OVLT of the 9-day-old rat, serotonin-containing terminals were observed to contact neurons and also to abut the perivascular space of capillaries (Ugramov et al. 1986). In the area postrema, there are relatively few serotonergic terminals (Fig. 21c). They have been observed electron microscopically (Pickel 1985).

The only sensory CVO in which serotonin-containing neurons are located within the body of the organ is the area postrema (Lanca and van der Kooy 1985; Newton et al. 1985). These serotonin-containing neurons have efferent connections to the lateral parabrachial nucleus (Lanca and van der Kooy 1985). In regard to the subfornical organ and OVLT, although serotonergic neurons are absent within these organs, many of the supraependymal neurons that are located on their ventricular surfaces in rats are serotonergic (Takeuchi and Sano 1983). Such supraependymal serotonergic neurons have also been detected on the surface of the subfornical organ of the house musk shrew (Hirunagi et al. 1995), but not on the surface of guinea pig CVOs (Tramu et al. 1983).

5.2.14 Vasopressin

The presence of vasopressin within neurons or fibres of sensory CVOs is dependent on species and possibly metabolic state. Neurophysin II is synthesised with vasopressin as part of the glycosylated preprohormone that is the vasopressin precursor molecule. Neurophysin/vasopressin immunoreactivity was observed in fibres in the subfornical organ of the rat (Sofroniew 1985) and the tree shrew *Tupaia* (Weindl and Sofroniew 1985). Alm et al. (1997) reported vasopressin-containing cell bodies in the subfornical organ of female rats.

In regard to the OVLT, it has been reported that either vasopressin- or neurophysin-containing fibres are present in the OVLT of the rat but not in the primate OVLT (Weindl and Sofroniew 1985). We were unable to detect vasopressin-containing fibres or cell bodies with immunohistochemical techniques in the OVLT of normal sheep; however, we did observe a profusion of vasopressin-containing fibres to enter the OVLT and occasionally the subfornical organ of sheep in which the hypothalamo-neurohypophysial pathway had been destroyed by ablating the median eminence (McKinley et al. 1996a). These vasopressin-containing fibres appeared to terminate in the intrapial primary plexus of the OVLT and to originate from the supraoptic and possibly paraventricular nucleus of the hypothalamus.

The area postrema of *Tupaia* was reported to receive the occasional vasopressin-(neurophysin) containing fibre from the adjacent dorsal vagal complex (Weindl and Sofroniew 1985). Alm et al. (1997) were unable to detect vasopressin immunoreactivity in rat area postrema.

6 The Neural Connections of the Sensory Circumventricular Organs

If the subfornical organ, OVLT and area postrema function as sensors, an essential component for this function will be neural pathways linking these sensory sites to integrative and effector motor regions of the brain. Several neuroanatomical studies have mapped direct efferent connections of the sensory CVOs. These studies have utilised tracer molecules that are transported by axons from their sites of injection either in the anterograde or retrograde direction relative to neuronal cell bodies. Electrophysiological studies have also yielded data that are relevant to the neuronal connectivity of the sensory CVOs.

6.1 Subfornical Organ

6.1.1 Efferent Neural Connections

The great majority of studies of the neural connections of the subfornical organ have been made in the laboratory rat. A number of efferent projections of neurons within the subfornical organ of this species are well recognised and are summarised in Table 3. Initial studies utilised autoradiography to trace the anterograde transport of tritiated leucine and proline that had been microinjected into the subfornical organ (Miselis et al. 1979). Later work employed horseradish peroxidase (HRP), wheat germ agglutinin conjugate of horseradish peroxidase (WGA-HRP), *Phaseolus vulgaris* leucoagglutinin (PHA-L) and biotinylated dextran as anterogradely transported markers. Confirmation of many of these efferent pathways has come from studies of retrogradely transported tracers (e.g. HRP, fast blue, fluorescent microspheres, cholera toxin B subunit) microinjected into proposed terminal fields.

The routes by which efferent fibres leave the subfornical organ have been described in detail by Miselis (1981) and Lind et al. (1982). Fibres travel mostly through its ventral stalk and into the adjacent median preoptic nucleus (MnPO). From there, fibres continue in a ventral direction through this nucleus predominantly around the rostral edge of the anterior commissure, although some also course around its caudal boundary close to the third ventricle. Fibres either leave this main group to proceed caudally and laterally into other preoptic and hypothalamic regions or continue in a ventral direction into the remainder of the MnPO and then to the OVLT, suprachiasmatic and supraoptic nuclei. In addition to this major midline fibre bundle, another group of fibres leaves the ventral stalk of the subfornical organ,

Table 3 Direct efferent and afferent neural connections of the subfornical organ

Efferent terminal fields	Origins of afferent inputs	References
Telencephalon		
Substantia innominata		Swanson and Lind 1986
Medial septal n.	Medial septal n.	Lind et al. 1982; Miselis et al. 1987b
Lamina terminalis		
Median preoptic n.	Median preoptic n.	Miselis et al. 1979; Lind et al. 1982; Saper and Levisohn 1983
OVLT	OVLT	Miselis et al. 1979; Oldfield et al. 1992; Gu and Simerly 1997; Miselis et al. 1987b
Limbic structures		
Bed n. stria terminalis	Bed n. stria terminalis	Swanson and Lind 1986; Miselis et al. 1987b
Central and median amygdaloid n.		Miselis et al. 1987b
Infralimbic cortex		Swanson and Lind 1986
Preoptic region		
Lateral preoptic area	Dorsal preoptic n.	Lind et al. 1982
Anteroventral periventricular n.	Anteroventral periventric. n.	Lind 1987
Medial preoptic area	Medial preoptic area	Miselis et al. 1987b; Lind et al. 1982
Preoptic periventricular n.		Miselis et al. 1987b
Hypothalamus		
Arcuate nucleus.	Anterior hypothalamic area	Lind et al. 1982
Arcuate n.		Rosas-Arellano et al. 1986; Miselis et al. 1987b; Krout et al. 2001.
Dorsal perifornical area	Dorsal perifornical area	Lind et al. 1984; Miselis et al. 1987
Dorsomedial nucleus	Dorsomedial n.	Swanson and Lind 1986; Miselis et al. 1987
Lateral hypothalamus	Lateral hypothalamus	Miselis 1981; Miselis et al. 1987b
Median eminence		Miselis et al. 1987b
Periventricular stratum		Miselis 1981
Paraventricular n.	Paraventricular n.	Miselis 1981; Larsen et al. 1991
Supraoptic n.		Miselis et al. 1979; Miselis 1981
Suprachiasmatic n.		Miselis et al. 1987b
Zona incerta	Zona incerta	Lind et al. 1984; Swanson and Lind 1986
Thalamus		
Centromedial n.		Miselis et al. 1987b
Paraventricular n.		Miselis et al. 1987b; Lind 1987
Reuniens n.	Reuniens n.	Lind et al. 1984; Miselis 1987b; Lind 1987
Midbrain		
	Locus coeruleus	Miselis et al. 1987b; Kawano and Masuko 2001
	Lateral parabrachial n.	Gu and Ju 1995, Lind 1987
Dorsal and median raphe		Lind 1987
Hindbrain		
	NTS	Zardetto-Smith and Grey 1987; Ciriello et al. 1986; Kawano and Masuko 2001
Ventrolateral medulla		Kawano and Masuko 2001

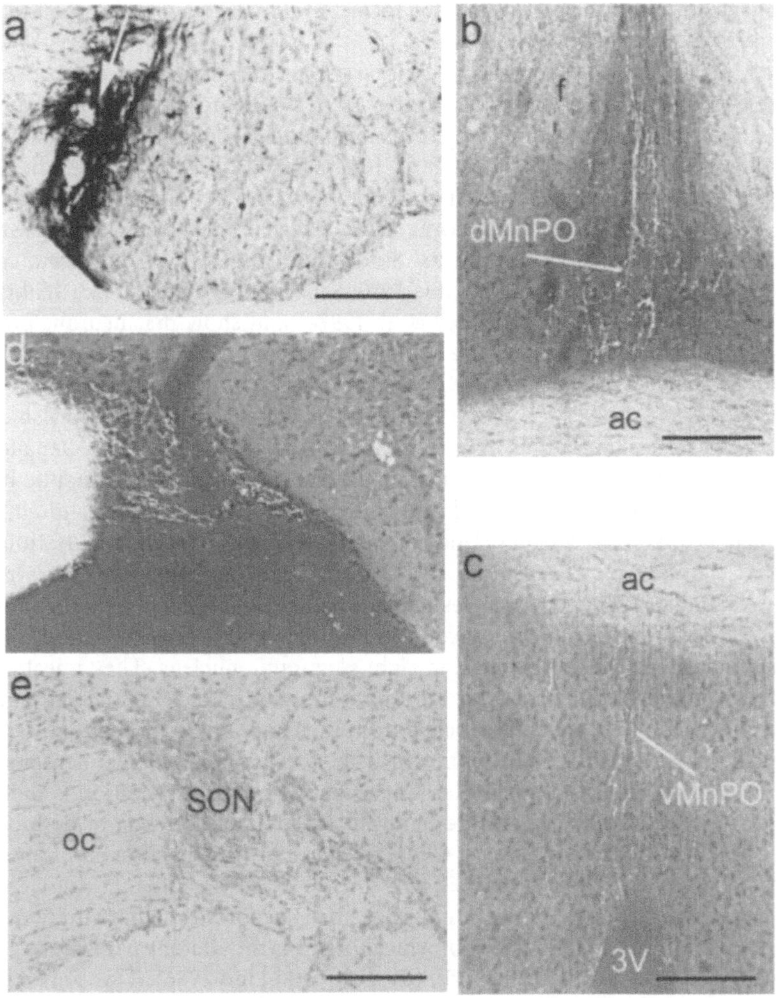

Fig. 22a–e Anterogradely transported biotinylated dextran amine (BDA) in coronal sections of the median preoptic and supraoptic nuclei after a unilateral microinjection of this tracer into the outer shell of the subfornical organ of a rat. **a** Coronal section of the subfornical organ showing the site of injection (*arrowed black region*) of BDA into the left side of the outer shell of the subfornical organ. A few retrogradely labelled neurons can be seen in the core of the subfornical organ. **b–d** Dark-field micrographs of coronal sections showing fibres of passage and terminals (white beaded fibres) in the median preoptic nucleus dorsal to the anterior commissure (**b**), ventral to the anterior commissure (**c**) and in the supraoptic nucleus (**d**). **e** Bright-field view of the section of the supraoptic nucleus shown in **d.** *ac*, Anterior commissure; *dMnPO*, dorsal median preoptic nucleus; *f*, fornix; *oc*, optic chiasm; *SON*, supraoptic nucleus; *vMnPO*, ventral median preoptic nucleus; *3V*, third ventricle. *Calibration bars*=150 μm

moving laterally into the columns of the fornix, eventually leaving this tract to terminate in a number of hypothalamic nuclei.

The MnPO is one of the major efferent targets of the subfornical organ. After microinjection of various anterogradely transported tracer molecules into the subfornical organ of the rat, labelled nerve terminals abound throughout the MnPO (Miselis et al. 1979, 1987; Miselis 1981; Lind et al. 1982) (Fig. 22b, c). In these studies, dense terminal fields were also observed in the supraoptic and hypothalamic paraventricular nuclei, and in the latter case, terminals were distributed throughout the magnocellular and parvocellular components. Studies utilising retrograde tracers confirm these projections (Tribollet and Dreifus 1980; Saper and Levisohn 1983; Wilkin et al. 1989; Anderson et al. 1990; McKinley et al. 1992c) and show that neurons predominantly in the peripheral parts of the subfornical organ give rise to the projections to the MnPO (Saper and Levisohn 1983), paraventricular nuclei and supraoptic nuclei in the rat (Miselis 1979; McKinley 1992c). Electrophysiological data are also consistent with these results (Ferguson et al. 1984; Tanaka et al. 1985, 1987; Renaud et al. 1985). Some neurons in the subfornical organ that project to the supraoptic nucleus have been shown to have collateral branching axons that also project to either the hypothalamic paraventricular or median preoptic nuclei (Weiss and Hatton 1990; Oldfield et al. 1991c). Because the subfornical organ (and the OVLT) are midline structures, Renaud et al. (1993) investigated whether individual neurons within these two CVOs might have branching axons so that they could provide neural inputs to both supraoptic nuclei i.e. the left and right supraoptic nucleus. They injected a different retrogradely transported tracer into the supraoptic nuclei on the left and right sides of the brain. Neurons were labelled from each supraoptic nucleus and remained essentially ipsilateral to the injection, with only a few double-labelled neurons showing collateral branches to both nuclei. Consistent with this observation is that retrogradely transported tracer from the paraventricular nucleus tends to be located in greater amounts on the side of the CVO ipsilateral to the injection site (Sunn et al. 2001).

The only other efferent projection that has been reported to come predominantly from neurons in the core of the subfornical organ is an efferent pathway to the bed nucleus of the stria terminalis (BNST) (Swanson and Lind 1986). This projection terminates in a restricted region of the rostral ventral BNST and adjacent parastrial nucleus (Swanson and Lind 1986; Miselis et al. 1987). We have confirmed this projection with the retrogradely transported tracer cholera toxin B subunit injected into the BNST (Sunn et al. 2003) and find that neurons in both the peripheral shell of the subfornical organ and its caudal inner core have efferent connections to the BNST and are activated by circulating angiotensin II (Sunn et al. 2003).

We have also obtained evidence in sheep, with the retrogradely transported tracer WGA-HRP, that efferent pathways from the subfornical organ to median preoptic and supraoptic nuclei exist in this species (McKinley et al. 1988; Oldfield et al. 1991c). Both neuroanatomical and electrophysiological studies show that as well as direct projections from the subfornical organ to both the supraoptic and paraventricular nuclei, indirect pathways to these two sites from the subfornical organ occur which are relayed via synapses in the MnPO (Tanaka et al. 1987; Oldfield et al. 1991a). A projection to the OVLT, predominantly to its dorsal cap, also is evident in the studies of Miselis et al. (1979) and Lind et al. (1982). It has been suggested that this projection arises from neurons in the central core of the subfornical organ

(Saper and Levisohn 1983); however, we have made minute injections of the anterogradely transported tracer biotinylated dextran into the outer shell of the subfornical organ (which did not overlap the ventromedial core of the subfornical organ) and observed terminal fields in the OVLT.

The subfornical organ also has efferent projections to the infralimbic part of the pre-frontal cortex, to the substantia innominata (ventral pallidium), rostral parts of the zona incerta, and lateral hypothalamus. The sources of these nerve terminalis are neurons in the periphery of the subfornical organ (Swanson and Lind 1986; Miselis 1981; Miselis et al. 1987). Other projections are to the dorsal perifornical area of the hypothalamus, lateral preoptic area and the preoptic and hypothalamic periventricular stratum, lateral preoptic area and the suprachiasmatic nucleus (Lind 1987; Krout 2001). Efferents to the midbrain raphe have also been identified (Lind 1987).

6.1.1.1
Viral Tracing

During the past decade, the neurotropic virus pseudorabies (PRV) has been used to map polysynaptic neural pathways from the CNS to peripheral organs. Transneuronal retrograde transport of the PRV after its injection into a number of peripheral organs (e.g. kidney, adrenal gland, sympathetic ganglia, salivary glands, heart, gut) reveals that the subfornical organ becomes infected approximately 95–100 h after the peripheral injection (Hubschle et al. 1998; Sly et al. 1999; Westerhaus and Loewy 1999; Yang et al. 1999). Invariably, it is neurons in the periphery of the subfornical organ, rather than its ventromedial core, that are labelled by the initial wave of viral infection (Fig. 23). These results indicate that the subfornical organ has polysynaptic pathways to the sympathetic and parasympathetic nerve supply to most, if not all, peripheral organs and tissues. It seems likely that synaptic relays via neurons in the PVN projecting to the intermediolateral cell column in the spinal cord and then to the sympathetic ganglia innervating particular organs may mediate influences of the subfornical organ on autonomic activity.

6.1.2
Afferent Neural Pathways of the Subfornical Organ

Earliest studies of the afferent connections with ablation techniques identified neural input to the subfornical organ coming from the neighbouring MnPO (Hernesniemi et al. 1972), and these have been confirmed by studies showing that tracers injected into the subfornical organ are retrogradely transported back to neuronal cell bodies in the median preoptic nucleus. Some of these neurons in the median preoptic nucleus projecting to the subfornical organ were also shown to have collateral branches to the supraoptic nucleus (Oldfield et al. 1992). In addition, there is evidence of forebrain afferent connections arising from the rostroventral part of the BNST, the dorsal cap of the OVLT and the anteroventral periventricular nucleus, suggesting reciprocal innervations (Miselis 1979; Lind et al. 1982; Miselis et al. 1987). In the diencephelon, neurons in the lateral hypothalamic region, zona incerta, arcuate nucleus, dorsome-

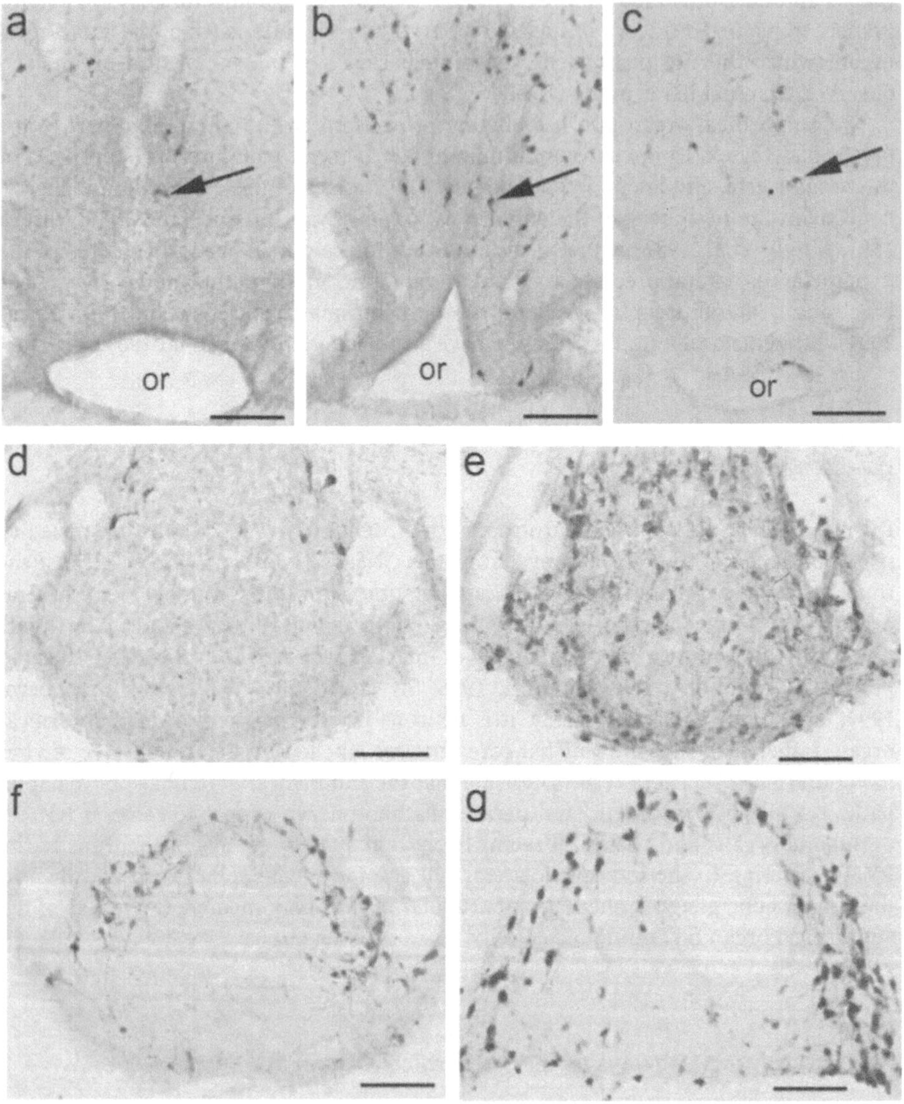

Fig. 23a–g Photomicrographs of coronal sections of the OVLT and subfornical organ of rats, showing retrogradely transported Bartha strain of pseudorabies virus (PRV), 96 h after its injection into the heart, forelimb skeletal muscle, kidney or submandibular gland. **a** OVLT and **d** subfornical organ after injection of PRV into the right atrium of the heart; **b** OVLT and **e** subfornical organ after injection of PRV into a forelimb muscle; **c** OVLT and **f** subfornical organ after injection of PRV into the kidney (see Sly et al. 1999 for details of methods). **g** Subfornical organ of a rat after injection of PRV into the submandibular gland (see Hubschle et al. 1998 for details). Retrogradely labelled neurons in the lateral part of the OVLT are indicated by *black arrows*. Viral labelling in the subfornical organ is confined to the outer shell. *or*, Optic recess of the third ventricle. *Calibration bars*=100 μm

dial hypothalamic nucleus and nucleus reuniens of the thalamus have neural projections to the subfornical organ (Lind 1987; Rosas-Arellano et al. 1996a,b). In the brain stem, afferent input to the subfornical organ comes from the dorsal and median raphe nucleus, latero-dorsal tegmental nucleus, periaqueductal grey, throughout the rostrocaudal extent of the lateral parabrachial nucleus, locus coeruleus, ventrolateral medulla and the nucleus of the solitary tract (NTS) (Lind 1987; Miselis et al. 1987; Zardetto-Smith and Gray 1988; Gu and Ju 1995; Kawano and Masuko 2001). This latter afferent pathway is mainly from noradrenergic neurons of the A2 cell group in the NTS (Ciriello et al. 1996b).

There is some disagreement in the literature regarding an afferent neural connection to the subfornical organ from the hypothalamic paraventricular nucleus. We and others were unable to find evidence for this connection after injections of WGA-HRP into the subfornical organ. However, evidence for this pathway was obtained after injection of PHA-L into the region of the paraventricular nucleus (Larsen et al. 1991).

It is possible that some afferent neural inputs to the subfornical organ have collateral branches to other related regions. For example, we observed that if different retrogradely transported tracers were injected into the subfornical organ and the supraoptic nucleus, retrogradely double-labelled neurons were observed in both the median preoptic nucleus and the OVLT. This is an indication of neurons in the OVLT and median preoptic nucleus that have collateral fibres branching to both the subfornical organ and supraoptic nucleus.

A polysynaptic pathway from the kidney to the subfornical organ, and many other central sites including the OVLT and area postrema, is suggested by the Fos induction in these sites that is observed after stimulation of afferent renal nerves (Solano-Flores et al. 1997). The activated neurons are located in the ventromedial core of the subfornical organ. The details of the synaptic links from the kidney remain to be determined.

6.2 Neural Connections of the OVLT

6.2.1 Efferent Projections

Studies of the neural connectivity of the OVLT are quite limited. However, several of the brain regions that are the target of efferent fibres coming from the OVLT have been unequivocally identified in rat brain. Most of the efferent fibres leaving the OVLT descend in the median forebrain bundle, whereas a few travel through the periventricular strata (Gu and Simerley 1997). Injections of anterogradely transported tracers (radiolabelled leucine and proline; biotinylated dextran) into the OVLT have resulted in the appearance of labelled terminals in the supraoptic nucleus and also in the perizonal nucleus outside the supraoptic nucleus (Phillips and Camacho 1987). In rats, we observed that if the injection site of anterograde tracer (biotinylated dextran) was located in the dorsal cap region of the OVLT, terminal labelling was considerably greater than if it was situated in more ventrolateral parts of the OVLT (Uschakov et al. 2001). This suggests that the dorsal cap region is the major source of

OVLT efferent connections to the supraoptic nucleus. This has been confirmed by several studies (McKinley et al. 1992c; Oldfield et al. 1994; Sunn et al. 2001) showing that tracer molecules injected into the supraoptic nuclei are retrogradely transported mainly to the dorsal cap of the OVLT (see Fig. 25a, b, Sect. 7). Studies in sheep also show a projection to the supraoptic nucleus (Oldfield et al. 1991).

A strong projection from the dorsal cap region of the OVLT to the hypothalamic paraventricular nucleus has also been observed in the rat, particularly to magnocellular regions (Gu and Simerley 1997; Uschakov et al. 2001), and this is confirmed by retrograde transport of cholera toxin B subunit injected into the PVN (Sunn et al. 2001). In addition, we observed that if the injection of this tracer was into more periventricular parvocellular parts of the PVN, the retrogradely labelled neurons in the OVLT were predominantly in its lateral parts rather than the dorsal cap regions (Sunn et al. 2001).

Neurons in the OVLT also send projections to the two other components of the lamina terminalis, the subfornical organ and the median preoptic nucleus in rats and sheep (Camacho and Phillips 1981; Oldfield et al. 1991c; Gu and Simerley 1997). Neural inputs from OVLT to the parastrial nucleus and adjacent BNST have also been observed, as well as efferent fibres distributing to the lateral preoptic and lateral hypothalamic region. We have observed neurons in all parts of the OVLT that were retrogradely labelled after injection of fluorogold into the lateral hypothalamic region (Uschakov et al. 2001). Terminal fields are found in the midbrain in the periaqueductal grey region and lateral parabrachial nucleus (Gu and Simerley 1997). Our studies with retrogradely transported tracers have confirmed these projections of neurons in the OVLT to midbrain regions. We also obtained evidence of a connection to the medullary raphe pallidus nucleus. Camacho and Phillips (1981) have observed ascending nerve fibres leaving the OVLT to spread through the septal area and terminate in the cingulate cortex and hippocampus, and we have confirmed a projection from the dorsal cap of the OVLT to the cerebral cortex of rat with retrogradely transported tracer injected into the cingulate gyrus (A. Uschakov, B.J. Oldfield and M.J. McKinley, unpublished observations).

6.2.1.1
Polysynaptic Efferent Pathways to the Periphery

Similar to the subfornical organ, neurons in the OVLT are labelled with pseudorabies virus 4–5 days after injection of the virus into several peripheral organs and sympathetic ganglia (Hubschle et al. 1998; Sly et al. 1999; Westerhaus and Loewy 1999). Unlike the subfornical organ, where the viral labelling was localised to its outer shell, in the OVLT neurons infected with the virus were located along the lateral periphery and dorsal cap of the organ (Fig. 23) and extended into adjacent preoptic regions. The synaptic steps in the pathway from the OVLT to the periphery remain to be clarified; however, similar to the subfornical organ, a link through the hypothalamic PVN seems likely. However, in view of the efferent projections of the OVLT to the lateral hypothalamic region, and the viral infection of lateral hypothalamic neurons that occurs after peripheral inoculations, it is possible that the hypothalamic relays from the OVLT to the periphery are more widespread than just those to the PVN. A poly-

Table 4 Direct efferent and afferent neural connections of the OVLT described in the literature

Efferent terminal fields	Origins of afferent inputs	References
Lamina terminalis		
Median preoptic n.	Median preoptic n.	Gu and Simerly 1997; Saper and Levisohn 1983
Subfornical organ	Subfornical organ	Miselis 1981; Lind et al. 1982, Gu and Simerly 1997; Oldfield et al. 1991
Limbic structures		
Cingulate cortex		Camacho and Phillips 1981
Bed n. stria terminalis		Uschakov et al. 2001
Preoptic region		
Parastrial nucleus		Gu and Simerly 1997
Lateral preoptic area		Camacho and Phillips 1981
Medial preoptic area	Medial preoptic area	Gu and Simerly 1997; Camacho and Phillips 1981; King et al. 1982
Hypothalamus		
Arcuate n.	Anterior hypothalamus	Gu and Simerly 1997; Camacho and Phillips 1981
Periventricular region		Gu and Simerly 1997
Lateral hypothalamus	Lateral hypothalamus	Uschakov et al. 2001; Camacho and Phillips 1981
Paraventricular n.	Paraventricular n.	Larsen et al. 1991; Gu and Simerly 1997; Sunn et al. 2001; Miselis et al. 1979; Wilkin et al. 1989; Oldfield et al. 1991, 1994
Supraoptic n.	Ventromedial nucleus	Camacho and Phillips 1981
Thalamus		
Paraventricular n.		Uschakov et al. 2001
Midbrain		
Periaqueductal grey	Periaqueductal grey	Gu and Simerly 1997; Camacho and Phillips 1981
Locus coeruleus		Camacho and Phillips 1981
Lateral parabrachial n.		Gu and Simerly 1997
Hindbrain		
Raphe pallidus		Uschakov et al. 2001

synaptic pathway from the OVLT to the hypothalamic suprachiasmatic nucleus has also been identified by means of viral tracing (Krout et al. 2001).

6.2.2 Afferent Projections

Neural input to the OVLT is summarised in Table 4. It includes contributions from other parts of the lamina terminalis, the subfornical organ and the median preoptic nucleus (Miselis et al. 1979; Saper and Levisohn 1983; Oldfield et al. 1992). Other hypothalamic sites that provide afferent neural input for the OVLT are the ventromedial, arcuate, paraventricular and anterior hypothalamic nuclei as well as the medial preoptic and posterior hypothalamic regions (Camacho and Phillips 1981). Inputs

Table 5 Direct efferent and afferent neural connections of the area postrema described in the literature

Efferent terminal fields	Origins of afferent inputs	References
Hypothalamus		
	Anterior dorsomedial n.	Shapiro and Miselis 1985
	Dorsal perifornical region	Shapiro and Miselis 1985
Paraventricular n	Shapiro and Miselis 1985; Miselis et al. 1987a	
Midbrain		
Lateral parabrachial n.	Lateral parabrachial n.	van der Kooy and Koda 1983; Shapiro and Miselis 1985; Herbert et al. 1990; Cunningham et al. 1994
Mesencephalic n. of trigeminal nerve		Shapiro and Miselis 1985
Periaqueductal grey		Shapiro and Miselis 1985
Pericentral dorsal tegmental n.		van der Kooy and Koda 1983
Dorsal tegmental n.		van der Kooy and Koda 1983
Cerebellum		
Cerebellar vermis		Shapiro and Miselis 1985
Hindbrain		
N. of the solitary tract	N. of the solitary tract	Shapiro and Miselis 1985; Strominger et al. 1994; Cunningham et al. 1994
Dorsal motor vagal n.		Shapiro and Miselis 1985
Ambiguus n.		Shapiro and Miselis 1985
Rostral ventrolateral medulla		Blessing et al. 1987; Badoer et al. 1994
Paratrigeminal n.		Shapiro and Miselis 1985
Spinal trigeminal n.		Shapiro and Miselis 1985

from the midbrain may come from the locus coeruleus and periaqueductal grey (Camacho and Phillips 1981). As mentioned above, afferent polysynaptic links from the kidney to the OVLT may exist.

6.3 Area Postrema

6.3.1 Efferent Neural Connections

Much of the knowledge of these efferent neural pathways is based on the neuroanatomical studies of van der Kooy and Koda (1983), Shapiro and Miselis (1985), Miselis et al. (1987), Herbert et al. (1994), Strominger (1994) and Cunningham et al. (1994), who utilised microinjections of WGA-HRP, CT-HRP or PHA-L into the area postrema. The direct efferent neural connections of the area postrema are summarised in Table 5. Nerve fibres leave the caudal region of the area postrema and proceed either caudally over the central canal in the commissural NTS or rostrolaterally in the most dorsomedial part of the NTS. Within the NTS, efferent fibres from the area postrema terminate in the commissural NTS, caudal to the level of the area postrema, and more rostrally in the dorsolateral, medial and commissural NTS, which receive the richest innervation (Shapiro and Miselis, 1985). Dense terminals were seen surrounding the A2 noradrenergic and particularly the C2 adrenergic groups of neurons in the NTS. This latter group was situated midway between the solitary tract and the area postrema and has been termed the C2d (dorsal strip) cell group (Cunningham et al. 1994). Within the dorsal vagal complex, few if any fibres terminate in the dorsal motor nucleus of the vagus. Miselis et al. (1987) made the point that the dendrites of vagal motor neurons extend into the overlying medial and commissural NTS, making it possible that these dendrites could receive neural inputs from the area postrema; however, later studies (Cunningham et al. 1994) show that this is unlikely. Within the medulla, a few laterally directed efferent fibres from the area postrema have been reported to terminate in the spinal trigeminal and paratrigeminal nuclei (Shapiro and Miselis 1985); however, later work was unable to confirm this projection (Cunningham et al. 1994).

The major efferent pathway from the area postrema that projects outside the dorsal vagal complex leaves the caudolateral area postrema to proceed under the solitary tract, then ventrolaterally through the medullary reticular formation to the A1 noradrenergic group of the caudal ventrolateral medulla and then to the nucleus ambiguus, where some fibres terminate (Shapiro and Miselis 1995). Although nerve fibres from the area postrema terminating in the ventrolateral medulla have been questioned by Cunningham et al. (1994), it has been shown that tracers (fluorogold or fluorescent microspheres) injected into the rostral ventrolateral medulla (RVLM) are retrogradely transported back to catecholaminergic neurons in the area postrema, confirming the efferent connection from area postrema to RVLM (Blessing et al. 1987; Badoer et al. 1994; Polson et al. 1996).

After passing through the caudal ventrolateral medulla, this group of fibres then turns rostrally to innervate the remainder of the nucleus ambiguus or ascend near the ventrolateral surface of the medulla, eventually turning dorsally in the lateral lemniscus at the level of the superior olivary complex to reach its major terminal field in the middle third of the lateral parabrachial nucleus (van der Kooy and Koda 1983; Shapiro and Miselis 1985; Cunningham et al. 1994). This projection to the lateral parabrachial nucleus arises from neurons in the caudal two-thirds of the area postrema. Tracing studies utilising retrogradely transported agents confirm this projection to the lateral parabrachial nucleus (Lanca and van der Kooy 1985; Herbert et al. 1991; Krukoff et al. 1992). They show two groups of neurons in the area postrema which project to different subnuclei of the lateral parabrachial nucleus. The first group of neurons in the medial part of the area postrema project to the waist region of the lateral parabrachial nucleus, the inner portion of the external subnucleus, central subnucleus and dorsal lateral subnucleus. The other group is in the lateral zone of the area postrema and has an ipsilateral efferent projection to the outer part of the external lateral subnucleus (Herbert et al. 1990). Shapiro and Miselis (1985) report a secondary terminal field rostrally in the lateral parabrachial nucleus. It has also been reported that cholecystokinin-containing and galanin-containing neurons in the area postrema have efferent connections to the lateral parabrachial nucleus (Herbert and

Saper 1990). The efferent neural outflow from the area postrema was also studied in ferrets, but no direct connections to the lateral parabrachial nuclei were observed and the only efferent pathway seen in this species was to the NTS (Strominger et al. 1994).

In the rat, fibres that leave the lateral parabrachial nucleus may then terminate in the ventrolateral aspect of the midbrain periaqueductal grey, the mesencephalic nucleus of the fifth nerve and unspecified parts of the cerebellum (Shapiro and Miselis 1985), although these terminal fields were not seen in rats in which PHA-L was used as the anterogradely transported tracer, and it has been suggested that some of these connections may be from the adjacent NTS (Cunningham et al. 1994).

Neural projections from the area postrema to the pericentral dorsal tegmental and dorsolateral tegmental nuclei were reported by van der Kooy and Koda (1983), but these could not be confirmed by Shapiro and Miselis (1985). A polysynaptic pathway from area postrema to the suprachiasmatic nucleus has been identified (Krout et al. 2001).

6.3.2 Afferent Projections

Unlike the two forebrain sensory CVOs, the area postrema receives direct inputs from visceral sensory nerves, as well as being able to monitor the composition of the circulation. These neural inputs have been studied in several species and arise from vagal and glossopharyngeal nerve fibres, conveying sensory information from the abdomen and thorax (Kalia and Mesulam 1980; Ciriello et al. 1981; Kalia and Sullivan 1982; Contreras et al. 1982; Chernicky et al. 1983).

A limited number of afferent neural inputs to the area postrema originate from within the CNS. The most prolific of these is from the hypothalamus, and it originates from a continuous group of neurons encompassing the lateral parvocellular division of the paraventricular nucleus, the anterior part of the dorsomedial hypothalamic nucleus, periventricular stratum and the perifornical region (Shapiro and Miselis 1985). A few neurons within the lateral parabrachial nucleus and NTS also send nerve fibres that terminate in the area postrema, and a polysynaptic neural input from the kidney may also exist (Solano-Flores et al. 1997).

7 Immediate-Early Gene Expression in Sensory Circumventricular Organs

During the past 15 years, study of expression of genes such as *c-fos* and *c-jun* has provided invaluable tools for studying the function of either individual cells or whole populations of neurons in the CNS. This approach has been especially helpful in studying the sensory circumventricular organs, where blood-borne stimuli that do not cross the blood-brain barrier interact with neurons in these regions. After stimulation of a neuron, either through its synaptic input or hormonal receptors or by transduction of stimuli related to its sensory role, second messenger pathways are engaged. These lead to increased intracellular calcium levels that subsequently result in the increased expression of a number of genes that include *c-fos*. Synthesis of its encoded protein product, Fos, follows. This protein forms heterodimers with other proteins, and these bind to AP1 sites in DNA to act as a regulator of gene transcription (Morgan and Curran 1989b). Activated neurons can thus be identified either by immunohistochemical techniques to detect Fos, the protein encoded by *c-fos*, or by in situ hybridisation identification of its mRNA (Morgan et al. 1987; Hunt et al. 1987; Dragunow and Faull 1989; Morgan and Curran 1989a, b; Ceccatelli et al. 1989; Hoffman et al. 1990).

We and others have examined the distribution of activated neurons in sensory CVOs after several different stimuli.

7.1 Angiotensin II

7.1.1 Intravenous Infusion of Angiotensin II

Many neurons throughout the subfornical organ and around the periphery of the OVLT express Fos in response to intravenous infusions of angiotensin II (Fig. 9) in rats and rabbits (McKinley et al. 1992a, 1995; Rowland et al. 1994; Potts et al. 1999). This also occurs in the area postrema in the rat and rabbit; however, it is likely that most, if not all, of the Fos immunoreactivity (Fos-IR) in this CVO is secondary to baroreceptor stimulation (McKinley et al. 1992a, Potts et al. 1999). We showed that intravenous infusions of phenylephrine at rates that increased arterial pressure by 40 mmHg did not stimulate Fos production in the subfornical organ or OVLT but caused similar levels of Fos-IR in the area postrema, as did intravenous infusion of angiotensin II. Potts et al. (1999) demonstrated that after intravenous infusion of angiotensin II, Fos expression in the area postrema of rabbits, but not the subfornical organ or OVLT, was prevented by surgical removal of baroreceptors in the carotid sinus and aorta. This result further supports our conclusion that most Fos-IR in the area postrema following systemic angiotensin II infusion is secondary to increased arterial pressure.

The blood levels of angiotensin II expected to result from intravenous infusions at 1–5 μg/h, which we used in early studies (McKinley et al. 1992a, 1995), would be in the high physiological to supraphysiological range (Mann et al. 1980; Johnson et al. 1981). If the dose of angiotensin II is reduced to 0.3 μg/h, which produces moderate blood levels, similar to those incurred during conditions such as sodium depletion or severe dehydration (Mann et al. 1980), Fos-IR in the subfornical organ is confined to its ventromedial core and that in the OVLT to its lateral margins and caudal periventricular tissues. Little if any Fos-IR is observed around the periphery of the subfornical organ or the dorsal cap region of the OVLT (McKinley et al. 1998).

The intravenous infusion of angiotensin II also results in increased numbers of Fos-IR neurons in a number of other brain regions e.g. hypothalamic supraoptic and paraventricular nuclei, bed nucleus of the stria terminalis, and nucleus of the solitary tract (McKinley et al. 1992a; Potts et al. 1999). In regard to the supraoptic nucleus, cholera toxin B subunit conjugated to gold (CTB-gold) was injected into this nucleus to be retrogradely transported back to CVOs in the lamina terminalis. A number of neurons in the periphery of the subfornical organ, and in the OVLT, which were labelled with CTB-gold also expressed Fos in response to intravenous infusion of angiotensin II (Oldfield et al. 1994). These results show the feasibility of angiotensin-sensitive neurons in the forebrain sensory CVOs driving vasopressin-secreting neurons in the supraoptic and paraventricular nuclei. Consistent with this, ablation of the subfornical organ prevented Fos-IR in the supraoptic and paraventricular nuclei in response to intravenous angiotensin II (Rowland et al. 1994).

In addition, we recently showed that a significant proportion of angiotensin-sensitive neurons in the core of the subfornical organ and lateral edges of the OVLT project to the bed nucleus of the stria terminalis (Sunn et al. 2003). However, the function of this angiotensin-stimulated pathway remains to be determined.

7.1.2 Intracerebroventricular Angiotensin II

Intracerebroventricular (ICV) infusion of angiotensin II also resulted in many neurons within the lamina terminalis increasing the expression of *c-fos* (Herbert et al. 1992). Infusion of angiotensin II at 1 ng/min for 90 min caused many neurons in the most periventricular caudal parts of the subfornical organ and OVLT, and throughout the median preoptic nucleus, to express Fos. The main difference to the result obtained with intravenous infusion of angiotensin II was the strong stimulation of median preoptic neurons by ICV angiotensin II, but not intravenous angiotensin II (McKinley et al. 1995). This may explain why ICV angiotensin II is a far more potent dipsogen than systemically infused angiotensin II, because there is considerable evidence to implicate the median preoptic nucleus in the regulation of vasopressin secretion and in the generation of thirst and water drinking behaviour (Mangiapane et al. 1983; Gardiner and Stricker 1985; Gardiner et al. 1985; Cunningham et al. 1992;

McKinley et al. 1999). Because the median preoptic nucleus contains high levels of angiotensin AT_1 receptors (Allen et al. 2000), it is likely that centrally administered angiotensin II directly stimulates neurons in this nucleus. The exclusion of blood-borne angiotensin II from this nucleus by the blood-brain barrier explains why intravenous angiotensin II failed to increase the number of Fos-labelled neurons in the median preoptic nucleus.

7.1.3 Endogenous Angiotensin II

A number of physiological conditions and pharmacological treatments are characterised by an increase in the secretion of renin by the kidney, with the subsequent generation in the blood of the decapeptide angiotensin I and then, through the action of angiotensin-converting enzyme (ACE), the active octapeptide angiotensin II. Increased circulating levels of endogenous angiotensin II may have actions on all the sensory CVOs, and Fos expression in their neurons has been investigated in conditions in which the circulating levels of angiotensin II increase, e.g. diuretic or peritoneal dialysis-induced sodium depletion, diet with a low sodium content, haemorrhage, isoproterenol treatment or polyethylene glycol-induced hypovolaemia.

7.1.3.1 Sodium Depletion

The concentration of angiotensin II in the blood increases during sodium depletion in the rat (Mann et al. 1980). Sodium depletion may result from treatment with the diuretic furosemide, peritoneal dialysis with an isosmotic sodium-free glucose dialysate solution, or bilateral adrenalectomy. In each of these conditions, the number of neurons exhibiting Fos-IR in the subfornical organ, OVLT and area postrema increases considerably (McKinley et al. 1992b; Han et al. 1995; Vivas et al. 1996; Rowland et al. 1996; McKinley et al. 1998; Weisinger et al. 2000). Subcutaneous injection of either an AT_1 receptor antagonist (e.g. losartan, ZD 7155, valsartan) or a high dose of an ACE inhibitor before sodium depletion prevents the increased Fos-IR in the subfornical organ and OVLT (Fig. 24), showing that it is due to elevated levels of endogenous angiotensin II in the circulation. The distribution of Fos -immunoreactive neurons in the OVLT in sodium-depleted rats is around the lateral and caudal parts of this CVO, with few labelled neurons in the dorsal cap, whereas that in the subfornical organ is predominantly in its ventromedial core (Fig. 24). These subregions of the OVLT and subfornical organ therefore are stimulated by angiotensin II under physiological conditions.

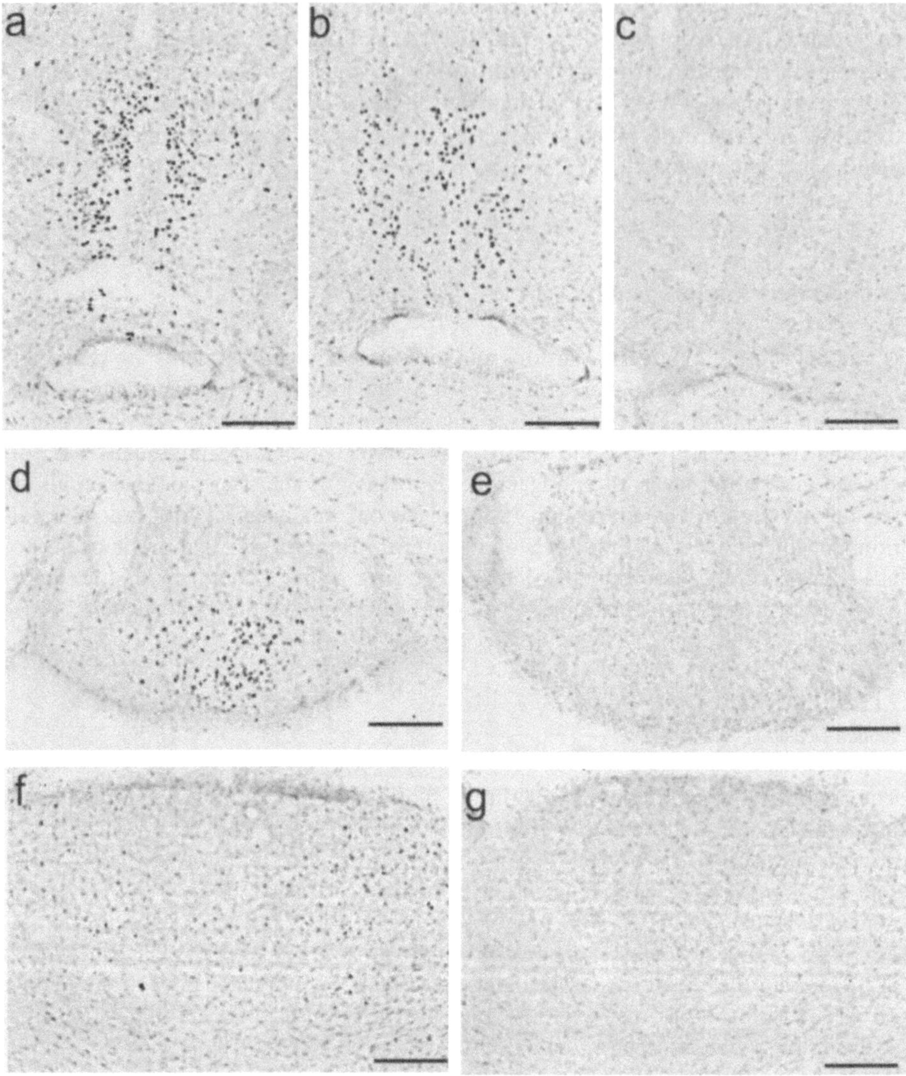

Fig. 24a–g Coronal sections of rat brain showing Fos-like immunoreactivity (Fos-IR), seen as black dots in cell nuclei in the OVLT (**a**, **b**, **c**), subfornical organ (**d**, **e**) or area postrema (**f**, **g**) 3 h after a subcutaneous injection of furosemide to induce sodium and water depletion (**a**, **b**, **d**, **f**) or with treatment with captopril (100 mg/kg s.c.) before such a furosemide injection (**c**, **e**, **g**). Fos-IR is located in the lateral and caudal periventricular parts of the OVLT (**a**, **b**), in the ventromedial core of the subfornical organ (**d**), and throughout the area postrema (**f**). Treatment with captopril did not inhibit the sodium and water losses in urine but prevented the Fos-IR in the sensory CVOs (**c**, **e**, **g**). *Calibration bars*=150 μm

7.1.3.2
Haemorrhage

We studied the effect of hypotensive and non-hypotensive haemorrhage in rats in a series of investigations (Badoer et al. 1992, 1993a, b). Withdrawal of blood from the circulation (either 2 or 4 ml) increased the number of neurons expressing *c-fos* in the subfornical organ, OVLT and area postrema, as well as several other sites in the brain (e.g. hypothalamic supraoptic and paraventricular nuclei, NTS and rostral and caudal ventrolateral medulla). The effect was far more robust when the haemorrhage was sufficient to cause a sustained reduction of arterial pressure, as when 4 ml of blood was withdrawn (Badoer et al. 1993a). The Fos-IR was observed throughout the subfornical organ and around the lateral and caudal parts of the OVLT and was blocked in these CVOs by pretreatment with the ACE inhibitor captopril, indicating that it was stimulated by endogenous circulating angiotensin II. This treatment did not block the *c-fos* expression in the area postrema, indicating that activation of neurons in this sensory CVO was probably due to neural inputs signalling the large fall (70 mmHg) in arterial pressure (Badoer et al. 1993b).

7.1.3.3
Isoproterenol

The β-adrenergic receptor agonist isoproterenol increases renin secretion as well as causing systemic vasodilatation and a fall in arterial pressure. Consequently, endogenous plasma angiotensin II levels are elevated considerably; this stimulates thirst. Subcutaneous injection of isoproterenol increases the number of neurons expressing Fos in the subfornical organ, OVLT and area postrema (Oldfield and McKinley 1994), as well as in a number of other brain regions. The number of Fos-labelled neurons in these regions is considerably greater than was observed in sodium-depleted rats. Prior peripheral administration of the AT_1 antagonist losartan, or the ACE inhibitor captopril, blocked this effect of isoproterenol and also blocked its dipsogenic response, showing that the expression of Fos in the CVOs and thirst were secondary to an action of angiotensin II (Oldfield and McKinley 1994; McKinley and Oldfield 1994).

7.1.3.4
Locally Generated Angiotensin II in the CVOs

ICV injection of renin causes drinking in rats and also stimulates the expression of Fos and another transcription factor, Egr-1, in several brain regions, including the subfornical organ. These responses are blocked by peripherally administered captopril (Xu and Johnson 1998). It seems likely that peripherally administered captopril acts outside the blood-brain barrier to prevent local formation of angiotensin II within the CVOs.

Another condition which may also result in local formation of angiotensin II in CVOs is treatment with relatively low doses of ACE inhibitors peripherally. We observed that subcutaneous administration of the ACE inhibitors captopril or enalapril

(0.5 mg/kg) caused water drinking in rats and resulted in the expression of Fos in many neurons throughout the subfornical organ and around the periphery of the OVLT (McKinley et al. 1997b). This dose of ACE inhibitor (0.5 mg/kg) is sufficient to block formation of angiotensin II in the periphery and to greatly reduce blood angiotensin II concentrations. The effects on Fos expression and the accompanying thirst were abolished by peripheral pretreatment with the angiotensin AT_1 antagonist ZD 7155, indicating that these responses were secondary to an action of angiotensin II on its receptors in the subfornical organ and OVLT. It was also prevented by using a higher dose of ACE inhibitor (50 mg/kg), which is sufficient to block the high concentrations of ACE in the CVOs (see Sect. 5). Our interpretation of these results is that the lower dose of ACE inhibitor, although blocking angiotensin II formation peripherally, leads to high concentrations of circulating angiotensin I. This high level of angiotensin I is caused by a lack of feedback inhibition by angiotensin II on renin secretion, so that renin levels increase and enzymatically generate angiotensin I. These high levels of angiotensin I reach the CVOs via the bloodstream and are converted to angiotensin II locally by the extremely high levels of ACE in the CVOs which need a 100-fold higher dose of captopril or enalapril for blockade.

7.2 Baroreceptor Activation

Arterial baroreceptors are activated when arterial pressure increases. Intravenous infusion of the α-adrenergic agonist phenylephrine elevates arterial pressure and stimulates the baroreceptor reflex (Dampney 1994). We mapped the distribution of neurons throughout the hypothalamus and medulla oblongata after intravenous infusions of phenylephrine that increased arterial pressure by 30–40 mmHg. The only sensory CVO to be activated was the area postrema (McKinley et al. 1992a; Badoer et al. 1994). A similar result was obtained in rabbits infused with phenylephrine (Li and Dampney 1994; Potts et al. 1997). Many neurons in the dorsal commissural, medial and dorsal subnuclei of the NTS and in the caudal ventrolateral medulla also exhibited Fos-IR in response to the pressor stimulus (Badoer et al. 1994; Potts et al. 1997). Thus the first two neuronal relays of the baroreceptor reflex were labelled. In these experiments, a retrogradely transported neuronal tracer (rhodamine-labelled microspheres) was injected into the rostral ventrolateral medulla at least 1 week before experiments, and neurons in the caudal ventrolateral medulla at the level of the obex were double-labelled with tracer and Fos, indicating that they were the inhibitory interneurons of the baroreceptor reflex. In addition, a few neurons in the area postrema also exhibited both labels (Badoer et al. 1994); however, their precise function remains to be determined. In the rabbit, sinoaortic denervation abolished the increased Fos expression in the area postrema, indicating that this response is mediated by baroreceptor activation (Potts et al. 1997).

7.3
Carbachol

ICV injection of the muscarinic receptor agonist, carbachol has been shown to increase Fos-IR in many regions of the brain, including the sensory CVOs (Rowland et al. 1994), and this effect is blocked by centrally administered atropine (Xu et al. 2001). Within the subfornical organ, the Fos labelling was restricted to the outer dorsolateral shell of this CVO, with few if any labelled neurons in its ventromedial core (Fig. 9). In the OVLT, although many neurons increased the expression of *c-fos* in response to ICV injection of carbachol, the most intense labelling was observed in the dorsal cap region, with an absence of any labelled cells in the central vascular zone (Fig. 9). These distributions of Fos labelling in the CVOs correspond with regions exhibiting the highest levels of staining for acetylcholinesterase (Meszaros et al. 1969; Lind 1987) and suggest that the carbachol may have diffused to these sites and acted directly on neurons in the CVOs. A significant number of neurons in the subfornical organ and OVLT, as well as the median preoptic nucleus, which are stimulated by ICV injection of carbachol have efferent neural connections to the supraoptic nucleus and may mediate the vasopressin secretion caused by centrally administered carbachol (Xu et al. 2001).

7.4
Anorectic Peptides—Cholecystokinin, GLP-1 and Amylin

Intravenous infusion of the peptide CCK-8S, a peptide with potent satiety properties, in rats increases Fos expression in the area postrema as well as the nucleus of the solitary tract (NTS), amygdala and oxytocin-containing neurons of the supraoptic and paraventricular nuclei. An antagonist to the CCK-A receptor subtype blocked such responses, whereas an antagonist that blocked CCK-B receptors did not (Luckman et al. 1993; Day et al. 1994). Within the area postrema, Fos expression in response to CCK was greatest in the caudal part of the CVO, and some of the Fos-positive neurons also exhibited tyrosine hydroxylase immunoreactivity, indicating that they were noradrenergic (Rinaman et al. 1993). The finding that bilateral subdiaphragmatic vagotomy suppressed CCK-induced expression of Fos in the area postrema, as well as the other central sites, indicated that CCK was acting on abdominal vagal afferents to activate neurons in the area postrema and elsewhere (Day et al. 1994; Li and Rowland 1995). Intraperitoneal injection of CCK-8 increased Fos expression in the area postrema and NTS of fasted mice (Wang et al. 1998), and intravenous administration of CCK in macaques has also been shown to increase the expression of Fos in these regions (Schreihofer et al. 1997). These latter authors suggested that CCK may utilise similar neural pathways in primates and rodents to stimulate behavioural, autonomic and neuroendocrine responses.

Intravenous administration of other anorectic peptides from the gastrointestinal tract such as GLP-1 or amylin, also stimulated the expression of Fos in the area postrema, as well as the NTS, lateral parabrachial nucleus and central nucleus of the amygdala (Rowland et al. 1997). Thus Fos-IR in the area postrema is a common observation following the administration of agents that inhibit appetite.

7.5
Hypertonicity

7.5.1
Intravenous Infusion of Hypertonic Solutions

Evidence that cerebral osmoreceptors responsible for osmoregulatory thirst and vasopressin secretion were located in regions of the brain which lacked a blood-brain barrier (McKinley et al. 1978; Thrasher et al. 1980), such as the OVLT and subfornical organ (McKinley et al. 1982; Thrasher et al. 1982), led us to study the effect of intravenous infusions of hypertonic saline, sucrose or urea on Fos expression in rats. In comparison to control infusions of isotonic saline, intravenous infusion of hypertonic NaCl or sucrose caused many neurons in the subfornical organ, OVLT and median preoptic nucleus (MnPO) to exhibit Fos-IR (Oldfield et al. 1991b). We also tested the response to intravenous infusion of concentrated urea, which may readily cross the cell membrane of the osmoreceptor (Verney 1947). It did not stimulate Fos expression in the CVOs (M.J. McKinley, unpublished data). The Fos expression in response to hypertonic saline was greatest in the dorsal cap region of the OVLT (McKinley et al. 1998). In the subfornical organ, fewer cells were labelled, and these were predominantly, although not exclusively, in the outer dorsolateral shell (McKinley et al. 1998). Infusion of all three hyperosmotic solutions increased Fos-IR in the supraoptic and paraventricular nuclei, although it was less pronounced in the case of the infusion of urea. Other investigators have also observed that administration of hypertonic saline to rats stimulates expression of Fos mRNA, Fos protein, or other inducible transcription factors such as nur 77, egr1, Jun and Krox 24 in the lamina terminalis and supraoptic and paraventricular nuclei (Sharp et al. 1991; Guldenaar et al.1992; Hamamura et al. 1992; Kovacs and Sawchenko 1994; Xu and Herbert 1998; Luckman 1997; Moellenhoff et al. 1998). The lack of response of the subfornical organ and OVLT to intravenous infusion of concentrated urea is consistent with the classic physiological observations that such an infusion does not stimulate the osmoreceptor subserving vasopressin secretion, whereas infusions of hypertonic NaCl or sucrose do (Verney 1947).

We proposed that osmoreceptors in the OVLT and to a lesser extent in the subfornical organ send efferent signals to the supraoptic and paraventricular nuclei to stimulate vasopressin secretion and thirst (McKinley et al. 1984). Consequently, the expression of Fos in response to intravenous infusion of hypertonic saline was investigated in rats in which prior injection of a retrogradely transported neuronal tracer had been made into the supraoptic or paraventricular nuclei, sites of magnocellular vasopressin-secreting neurons. It was observed that up to 70% of neurons in the dorsal cap of the OVLT which had efferent connections to the supraoptic nucleus were stimulated by intravenous infusion of hypertonic saline (Oldfield et al. 1994). Neurons in the dorsal cap of the OVLT which project to the paraventricular nucleus also were activated by intravenous infusion (Fig. 25g) or intraperitoneal injection (Larsen and Mikkelsen 1995) of hypertonic saline. Retrogradely labelled neurons in the MnPO and the subfornical organ (Fig. 25d, e) also increased expression of Fos in response to systemic hypertonicity, albeit to a lesser extent than the OVLT (Oldfield et al. 1994; Larsen and Mikkelsen 1995). We consider that these osmosensitive neurons

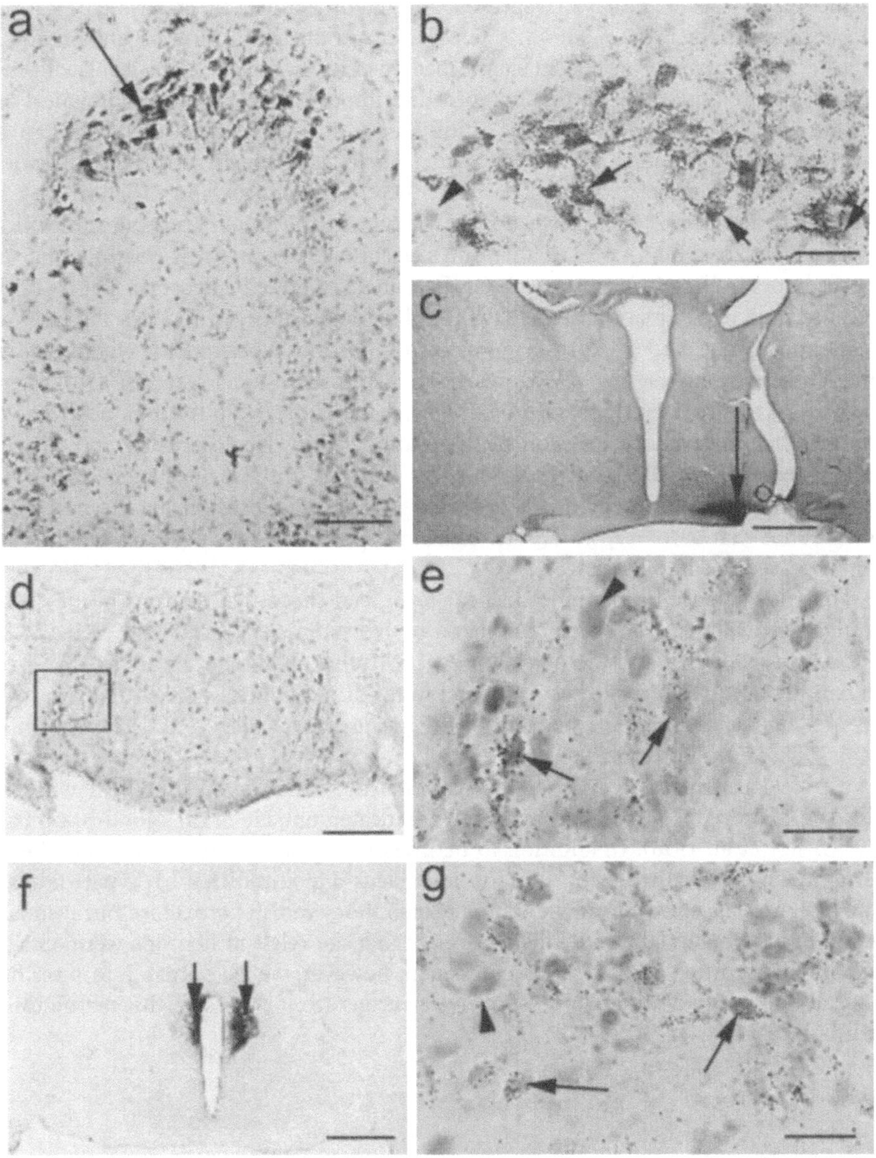

Fig. 25a–g Coronal sections of rat brain showing Fos-like immunoreactivity (brown cell nuclei) in neurons exhibiting cholera toxin B subunit-gold conjugate (CTB-gold) retrogradely transported from the supraoptic nucleus (**a**, **b**) or paraventricular nucleus (**d**, **e**, **g**). Rats had been infused with hypertonic saline intravenously 2 h before being killed. **a**, **b** Double-labelled neurons in the dorsal cap of the OVLT (*arrow* in **a**, low magnification). **c** Site of injection of CTB-gold into the supraoptic nucleus *(black arrow)*. **d**, **e** Double-labelled neurons in the subfornical organ, low magnification (**d**) and high magnification (**e**). **g** Double-labelled neurons in the dorsal cap of the OVLT. **f** Sites of injection (*arrows*) into the paraventricular nuclei. (Reproduced with permission from Oldfield et al. 1994). *Calibration bars*=50 μm in **a**, 200 μm in **d**, 0.8 mm in **c** and **f**

in the dorsal cap of the OVLT represent the major population of physiological osmoreceptors in the brain, with other populations in the subfornical organ and possibly the supraoptic, paraventricular and median preoptic nuclei. Consistent with this conclusion are data showing that combined lesion of both OVLT and subfornical organ, or unilateral lesion of the side of the MnPO to sever connections from lamina terminalis to supraoptic and paraventricular nuclei, severely disrupts Fos expression in these latter two nuclei (Xu and Herbert 1996; Hochstenbach and Ciriello 1996). In the case of the MnPO, it receives strong afferent neural input from the subfornical organ (Fig. 22b, c) and OVLT (Saper and Levisohn 1983), and it is possible that osmoreceptors in these two sensory CVOs drive median preoptic neurons. This is likely because individual ablation of the OVLT and subfornical organ reduces the number of neurons in the MnPO expressing Fos in response to hypertonicity, whereas combined ablation of these two CVOs nearly abolishes such a Fos response (Hochstenbach and Ciriello 1996). Increased Fos-IR has been observed in the area postrema of the rat after intravenous infusion of hypertonic saline (Hochstenbach and Ciriello 1996), and it is possible that osmosensitive neurons are also located in this CVO. However, because arterial pressure increases with such an infusion, an alternative explanation is that baroreceptor inputs to the area postrema have been activated, causing increased Fos in this sensory CVO.

There is evidence from studies in near-term fetal sheep that neurons in the OVLT and subfornical organ are also stimulated by hypertonicity (McDonald et al. 1998). This is consistent with physiological observations that ablation of these two CVOs results in reduced vasopressin secretion and water drinking in response to intracarotid or intravenous infusion of hypertonic saline in sheep (McKinley et al. 1984; McKinley et al. 1988; McKinley et al. 1999).

We have also made a study of the osmosensitive neurons in the OVLT with the electron microscope to confirm that they are located outside of the blood-brain barrier (Bisley et al. 1996). In addition to neurons in the dorsal cap of the OVLT being retrogradely labelled from the supraoptic nucleus and stimulated by hypertonic saline to express Fos, we administered the protein horseradish peroxidase intravenously 5 min before animals were killed to verify that the relevant neurons were outside the blood–brain barrier. This was confirmed; however, we were unable to morphologically differentiate the osmoresponsive neurons from others in the surrounding neuropil.

7.5.2 Dehydration

Rats were dehydrated by depriving them of water for 24 or 48 h. Subsequent immunohistochemical detection of Fos in the brain showed that there were many neurons which increased Fos expression throughout the lamina terminalis, particularly in the OVLT, median preoptic, supraoptic and paraventricular nuclei (McKinley et al. 1994a; Xu and Herbert 1996; Patronas et al. 1998; Morien et al. 1999). Fos expression in the subfornical organ, although increased, was less intense than in the other two regions (McKinley et al. 1994a). As expected, both supraoptic and paraventricular nuclei also exhibited intense Fos-IR in these dehydrated rats. Unlike rats infused with hypertonic saline, the Fos-IR in the OVLT was not confined mostly to the dorsal cap region but was also present throughout the organ except for the central vascular zone. This probably reflects the fact that depletion of both intracellular and extracellular volume occurs in dehydrated animals, whereas only the intracellular compartment is dehydrated with intravenous infusion of hypertonic saline. The additional depletion of the extracellular fluid in dehydrated animals will stimulate renin secretion, with subsequent increase in blood angiotensin II levels that will cause Fos expression in the other parts of CVOs (see Sect. 7.1).

We also observed that many neurons in the dorsal cap of the OVLT and median preoptic nucleus (approximately 30%), and fewer neurons in the subfornical organ (6%) which were retrogradely labelled with cholera toxin B subunit-gold conjugate that had previously been injected into the supraoptic nucleus expressed Fos in response to water deprivation (McKinley et al. 1994a). These neurons probably drive neurons in the supraoptic nucleus to secrete vasopressin in response to dehydration.

7.6 Lipopolysaccharide (Endotoxin)

Intravenous or intraperitoneal injection of the bacterial endotoxin lipopolysaccharide (LPS) results in increased Fos expression in several regions of the brain, including the three sensory CVOs (Elmquist et al. 1993; Sagar et al. 1995; Lacroix and Rivest 1997). This Fos immunoreactivity is spread throughout each of the CVOs and is also apparent along their meningeal and ventricular surfaces, indicating that the LPS influences glial and ependymal cells as well as neural perikarya. Of relevance to this is the observation that interleukin-1 administration also stimulates Fos expression in the area postrema and OVLT (Ericsson et al. 1994). Elmquist et al. (1996) studied different doses of LPS (5 and 125 μg/kg) and different periods of time after endotoxin administration. They observed that after 2 h, only the higher dose of LPS caused Fos expression in the sensory CVOs, whereas other regions such as the paraventricular nucleus, locus coeruleus, lateral parabrachial nucleus and NTS were activated by the lower dose. The Fos expression with the higher dose was predominantly in the dorsal cap of the OVLT and in the ventromedial core of the subfornical organ. Sometimes there was Fos expression in the meninges of the OVLT with the lower dose. The authors suggested that besides stimulation of production of cytokines and prostaglandins in the CVOs, there may be other ways by which endotoxin signals to the brain to induce fever and immune responses. These include the actions of LPS on vagal afferent nerves to transmit signals to the CNS to cause fever and the production of prostaglandin in the walls of penetrating blood vessels in the ventrolateral medulla. The lipophilic prostaglandin may then pass into the interstitium of the RVLM, where it may activate neural pathways projecting to the forebrain which mediate febrile and neuroendocrine responses.

The pathophysiological relevance of these results is discussed in Sect. 8.4.

7.7 Relaxin

Relaxin is a hormone secreted during pregnancy. It is so named for its property of relaxing the pelvic ligaments to facilitate the birth process. Relaxin, a polypeptide molecule, also has influence on the brain (Geddes and Summerlee 1995) via actions in the forebrain CVOs. Intravenous infusion of relaxin stimulates the expression of Fos in both the subfornical organ and OVLT (McKinley et al. 1998; Sunn et al. 2002). A characteristic distribution of Fos-IR following intravenous relaxin is observed in both of these CVOs (Fig. 9). Many Fos-labelled neurons are observed throughout the periphery of the subfornical organ, extending from its ventral stalk, making an outer shell of activated neurons. Few if any neurons in the ventromedial core of the subfornical organ exhibit Fos-IR. In the OVLT, the Fos-IR in response to intravenous relaxin is confined mainly to the dorsal cap, throughout its rostrocaudal extent (Fig. 9a, b). Only a few immunoreactive neurons are observed along its lateral margins or in the inner vascularised plexus of the OVLT. No Fos-IR was observed in the area postrema. In addition to activation of these neuronal populations in the CVOs, Fos-IR increased in the supraoptic nucleus and magnocellular parts (both oxytocin and vasopressin cells) of the hypothalamic paraventricular nucleus.

It is probable that intravenously administered relaxin acts directly on neurons in both the subfornical organ and the OVLT. If relaxin is infused intravenously into a rat in which the subfornical organ has been ablated, Fos-IR is still observed in the OVLT. Similarly, if the OVLT is ablated, relaxin administration results in Fos labelling of neurons in the subfornical organ. Moreover, relaxin directly stimulates increased electrical activity of neurons in isolated slices of subfornical organ in vitro (Sunn et al. 2002).

We have also shown recently that many neurons in the subfornical organ and OVLT, with direct neural projections to the supraoptic and paraventricular nuclei, express Fos in response to intravenous infusion of relaxin (Sunn et al. 2001). It is likely that these neurons drive vasopressin secretion in response to relaxin, and this may contribute to the reduced plasma osmolality that is characteristic of pregnancy in rats and humans.

7.8 Vomiting and Emetic Stimuli

The area postrema has long been regarded as a chemoreceptor trigger zone (Borrison and Brizzee 1951). The pattern of Fos expression in response to emetic stimuli has been studied in the brains of animals which have a vomiting reflex, such as ferrets and cats (not rats). In ferrets, intraperitoneal administration of cisplatin, a cancer chemotherapy agent, caused Fos expression throughout the area postrema and in the dorsomedial subnucleus of the NTS. The most dense Fos expression was observed in the lateral parts of the area postrema. It also stimulated retching and vomiting. Unilateral cervical vagotomy, or treatment with gransetron, a 5-HT_3 antagonist, stopped vomiting, and attenuated the cisplatin-induced Fos expression in the NTS, but had little if any effect on the expression of Fos in the area postrema. The authors concluded that cisplatin had a direct action on the area postrema, whereas the effect on the NTS was the result of stimulation of vagal afferent nerves (Reynolds et al. 1991).

Another emetic stimulus, intraduodenal administration of hypertonic saline to the anaesthetised ferret, also induced Fos expression in the area postrema and medial subnucleus of the NTS (Boissonade et al. 1994). Cats, some decerebrated and some intact, were induced to vomit by an initial dose of cisplatin followed by combinations of apomorphine, lobeline sulphate, naloxone and protoveratrine. They were killed 4–5 h after the initial emetic stimulation, and Fos expression was mapped in the brain. Miller and Ruggiero (1994) considered that the entire emetic pathway was labelled with Fos. In both groups, the labelled areas included the area postrema as well as the orosensory and commissural subnuclei of the NTS, which were the sensory components of the pathway. Premotor neurons in the retroambiguus nucleus and motor neurons of the phrenic nerve and dorsal motor nucleus of the vagus also exhibited dense expression of Fos (Miller and Ruggiero 1994). By contrast, other emetic agents such as lobeline or apomorphine resulted in Fos expression in the NTS but little if any in the area postrema (Miller and Leslie 1994).

It is clear from these studies that the area postrema does not have to be activated for all emetic stimuli to produce nausea and vomiting, and this conclusion has also been drawn from studies of lesions of the area postrema (for review, see Miller and Leslie 1994). However, it does appear to play a crucial chemoreceptor role, along with vagal afferent input to the NTS, for many emetics to induce vomiting.

8 Functions of the Sensory Circumventricular Organs

The morphology, neural connectivity, neurochemistry and modified barrier properties of the sensory CVOs dealt with in the preceding sections are guideposts to the functions that these regions of the CNS subserve. Investigations over the past 30 years have greatly enriched our understanding of why these CVOs are built the way they are and what their physiological and pathophysiological roles may be. High on the list are sensory functions related to body fluid regulation and cardiovascular control, but clearly the sensory CVOs may also have other important roles such as that of the area postrema in vomiting and the participation of the sensory CVOs in fever. Neurosecretion into the circulation may also be a function, particularly in regard to the OVLT, where LHRH-containing neurons may release this peptide into the bloodstream (Samson et al. 1980), possibly in a cyclic manner (Wenger and Leonardelli 1980; Wenger et al. 1981; Piva et al. 1982). There is no evidence of a portal secretion from the OVLT to the anterior pituitary, and the target for this LHRH released from the OVLT is unknown (Lescure et al. 1978). In this final section we review sensory functions of the CVOs.

8.1 Osmoregulation

A major sensory function of the OVLT and subfornical organ, and possibly also of the area postrema, is that of detecting changes in the tonicity of the extracellular fluid and initiating the appropriate compensatory homeostatic responses. That is, the OVLT and subfornical organ house physiological osmoreceptors. In particular, osmoreceptors in the OVLT and subfornical organ initiate responses such as increased vasopressin secretion from the posterior lobe of the pituitary gland and thirst in response to increases in plasma osmolality.

The main evidence backing up these assertions comes from studies of the effects of lesions, neuroanatomical connections, electrophysiological characteristics and immediate-early gene (i.e. *c-fos*) expression. Initially the concept of the osmoreceptor as it pertained to the secretion of the antidiuretic hormone vasopressin was put forward by E. B. Verney, although Mayer had proposed such an idea for the stimulation of thirst in 1900. Verney was able to show in dogs that when the concentration of molecules that had slow passage across the cell membrane increased in the blood circulating to the brain, an antidiuresis ensued (Verney 1947), and this has been verified subsequently in other mammals (Olsson 1972; Ericsson et al. 1971; Blaine et al. 1975; Swaminithan 1980). Verney proposed that osmoreceptive neurons in the region of the hypothalamus responded to the effective osmotic pressure (i.e. tonicity) of blood, rather than absolute osmotic pressure. Thus, when the osmolality of blood is increased by addition of molecules such as NaCl or sucrose, osmoreceptor cells are dehydrated by osmosis because such polar molecules do not cross the semipermeable membrane and provide an effective osmotic gradient for water to move out of the cell. If smaller molecules such as urea or glycerol increase plasma tonicity, they diffuse rapidly into cells, do not osmotically dehydrate them and do not stimulate vasopressin release. Thus intracellular dehydration was considered to be the modus operandi of the osmoreceptor, with subsequent changes in the excitability of the osmoreceptor neuron resulting from the changed intracellular volume. Similar considerations were also made in regard to osmoreceptors that initiated thirst (Gilman 1937; Holmes and Gregersen 1950; Fitzsimons 1961; McKinley et al. 1978; Thrasher et al. 1980).

8.1.1 Sites of Cerebral Osmoreceptors

Jewell and Verney (1957) ligated various intradural branches of the internal carotid artery with the purpose of isolating the osmoreceptive region of the brain. They then tested the effects of intracarotid infusions of hypertonic saline on antidiuresis. They were able to localise the site of the osmoreceptors to a region which encompassed preoptic and hypothalamic tissue which also included the supraoptic and paraventricular nuclei of the hypothalamus, the site of the vasopressin-containing neurons that send axons to the posterior pituitary gland for releasing vasopressin into the circulation. Over time it became generally accepted that the neurosecretory neurons of the supraoptic and paraventricular nuclei were in fact the osmoreceptors. However, Jewell and Verney pointed out that other nearby regions 'the medial preoptic area, the suprachiasmatic nucleus, the nucleus supraopticus diffusis, anterior hypothalamic area, the periventricular system and the ventromedial nucleus must for the present be regarded as not inferior to those of the supraoptic nucleus itself' with respect to being the site of osmoreceptors (Jewell and Verney 1957). Another early study (Woods et al. 1966) also showed that the osmoreceptor regulation of vasopressin secretion persisted in decerebrate dogs, provided that an island of preoptic/hypothalamic tissue was maintained in these animals. If the general osmoreceptive regions outlined in these two studies are examined, it can be seen that a sensory CVO, the OVLT, is included in the crucial tissue. Other studies in which the effects of microinjections of hypertonic solutions into various diencephalic sites were studied (Peck and Blass 1975) also challenged the generally held view that magnocellular neurons of the supraoptic and paraventricular nuclei of the hypothalamus were the osmoreceptors regulating vasopressin secretion.

On reviewing these early studies, it appears that possible influences of the blood-brain barrier to exclude osmotically active substances from the preoptic/hypothalamic region have been overlooked (Ericsson et al. 1971). This led us to make intracarotid infusions of hypertonic solutions of NaCl, sucrose and urea in conscious sheep and to measure effects on water intake, renal free water clearance and the sodium concentration of the cerebrospinal fluid (McKinley et al. 1978). These studies showed that although the intracarotid infusions of hypertonic NaCl and sucrose were far more effective than hyperosmolar urea for the induction of water drinking and vasopressin secretion, the infusion of urea was nevertheless as effective as NaCl or sucrose at increasing the sodium concentration of the CSF. This effect of urea was attributed to its lack of passage across the blood-brain and blood-CSF barriers (Oldendorf 1971; McKinley et al. 1978), creating an effective osmotic gradient between blood and brain which results in withdrawal of water from the brain by osmosis. Therefore, it became evident that systemic infusions of concentrated urea solutions could be as effective as infusions of hyperosmolar NaCl or sucrose at osmotically dehydrating the brain but were relatively ineffective at stimulating the osmoreceptors. For such a sensor to distinguish urea from the other two osmotic agents, it was proposed that the cerebral osmoreceptors (at least in part) would need to be located in a brain region lacking a blood–brain barrier, such as a CVO (McKinley et al. 1978; Thrasher et al. 1980).

Attention became focused on the OVLT as a possible site of the osmoreceptors mediating vasopressin release and thirst. Ablation of tissue in the anterior ventral wall of the third ventricle, which included the OVLT along with the ventral median preoptic nucleus, anteroventral periventricular nucleus, medial preoptic and periventricular preoptic region, was shown to disrupt osmoregulatory vasopressin secretion and drinking in goats and rats (Andersson et al. 1975; Rundgren and Fyhrquist 1978; Johnson and Buggy 1978). These data were consistent with the idea that osmoreceptors were located within, or close to, the OVLT. More discrete lesions of the OVLT were shown to disrupt osmotically stimulated vasopressin secretion and water drinking in sheep and dogs (McKinley et al. 1982, 1984, 1985; Thrasher et al. 1982, 1987), and isolated hypothalamic explants from rats were found to secrete vasopressin into the bathing medium only if the OVLT region was included in the explant (Sladek and Johnson 1983). However, it also became clear from these studies that ablation of the OVLT did not totally disrupt vasopressin secretion and water drinking, suggesting that osmoreceptors existed in other brain regions as well.

More recent studies of the effects of lesions of various parts of the lamina terminalis indicate that ablation of the OVLT, in combination with lesions of the subfornical organ and/or median preoptic nucleus, which is located in the lamina terminalis between the two sensory CVOs, causes greater disruption of osmotically stimulated AVP release and thirst (McKinley 1985; McKinley et al. 1988, 1999) and largely prevents expression of *c-fos* in the supraoptic and paraventricular nuclei (Hochstenbach and Ciriello 1996). These results indicate the possibility of osmoreceptors or related pathways throughout all regions of the lamina terminalis. The neuroanatomical studies detailed in Sect. 5 indicate that direct neural projections proceed from all parts of the lamina terminalis to the supraoptic and paraventricular nuclei. In addition, pathways from the sensory CVOs may also relay through a synapse in the median preoptic nucleus (Oldfield et al. 1991a), from which direct connections to vasopressin-secreting neurons in the supraoptic nucleus have been demonstrated (Oldfield et al. 1990c). Electrophysiological studies both in vivo and in vitro indicate that neurons in all three subregions of the lamina terminalis (OVLT, subfornical organ and MnPO) respond with increased firing rates when exposed to physiological increases in tonicity (Sayer et al. 1984; Buranarugsa et al. 1979; Sibbald et al. 1988; Vivas et al. 1991; Nissen et al. 1993; McAllen et al. 1990; Gutman et al. 1988; Bourque et al. 1994; Renaud et al. 1993).

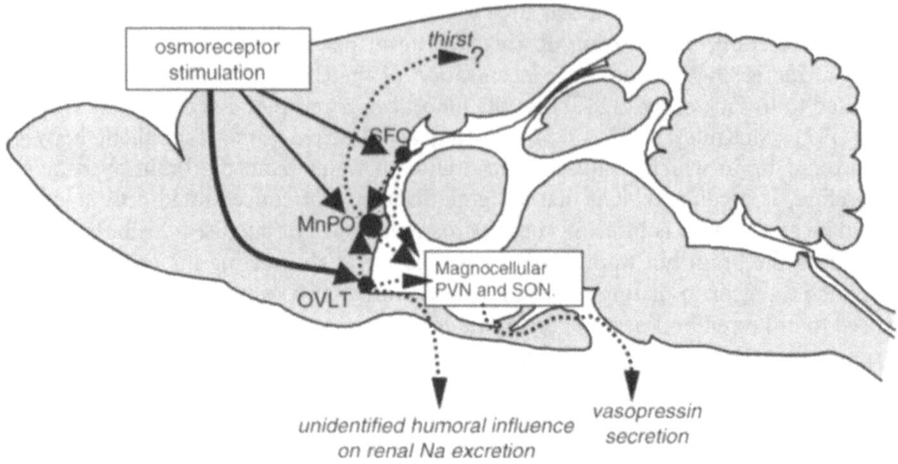

Fig. 26 Diagram of neural pathways in the rat brain that may mediate vasopressin secretion, thirst and natriuresis following stimulation of osmoreceptors in the circumventricular organs of the lamina terminalis and median preoptic nucleus. The cortical regions mediating thirst are unknown, and this is indicated by the *question mark*. *MnPO*, Median preoptic nucleus; *OVLT*, organum vasculosum of the lamina terminalis; *PVN*, hypothalamic paraventricular nucleus; *SFO*, subfornical organ; *SON*, supraoptic nucleus

Other electrophysiological studies in vitro, however, have indicated that the magnocellular neurons of the rat supraoptic nucleus may also be directly sensitive to increased tonicity of the extracellular fluid (Mason 1980). On the basis of these findings it has been proposed that the neural input to the supraoptic nucleus from the anteroventral third ventricle region (AV3V) is only providing background synaptic drive for the neurosecretory neurons of the supraoptic nucleus to express their innate osmosensitivity (Leng et al. 1989). However, the electrophysiological studies of Richard and Bourque (1992, 1995) and Nissen et al. (1993) show that it is unlikely that the drive to the supraoptic nucleus is merely background in nature; the OVLT provides osmotically sensitive drive to the neurosecretory neurons for vasopressin secretion (Richard and Bourque 1992, 1995). The osmotically stimulated expression of Fos in many neurons of the lamina terminalis that project to the supraoptic and paraventricular nuclei, and particularly those in the dorsal cap of the OVLT in the rat (Fig. 25a, b, g), is indicative of an osmoreceptor regulating vasopressin secretion. A diagram of the efferent neural pathways from the CVOs that may mediate osmoregulation is shown in Fig. 26.

Recently, a vanilloid receptor-related cation-selective channel, which is gated by hypotonicity, has been located by in situ hybridisation histochemistry in the OVLT, subfornical organ and median preoptic nucleus. It has been proposed that such an osmotically regulated channel could be the basis of an osmoreceptor in these CVOs (Liedtke et al. 2000). The clock gene *mPer3* is expressed in the OVLT in a circadian rhythmic fashion, and it has been suggested that it may influence rhythms of osmoreceptor function, LHRH secretion or temperature regulation (Takumi et al. 1998).

8.1.2
Osmoregulation of Natriuresis and Renin Secretion

In addition to osmoreceptors regulating thirst and vasopressin secretion, there is evidence that renal sodium excretion is also strongly influenced by a cerebral osmoreceptor (Blaine et al. 1975; McKinley et al. 1983c; Schloorlemmer 1999). This osmoregulatory control of sodium excretion results in a dehydration-induced natriuresis in several species (McKinley et al. 1983b). This has the effect of unloading electrolyte, thus ameliorating the increase in plasma osmolality. It is likely that the osmoreceptors regulating renal sodium excretion are located in the lamina terminalis, possibly in the OVLT (McKinley et al. 1992d), because ablation of the ventral lamina terminalis disrupts osmoregulatory natriuresis in sheep. Renal denervation does not disrupt osmoregulatory natriuresis (Park et al. 1989), and it is likely that this natriuretic response is mediated by a circulating hormone. It is possible that oxytocin may be such a hormone in the rat (Huang et al. 1995); however, this is not the case in other species, in which the humoral agent has yet to be identified (McKinley et al. 2000).

A central osmoreceptor located in the lamina terminalis may have an inhibitory influence on renin secretion (McKinley et al. 1986b, 1994d) which, by contrast with the osmoregulatory natriuresis, appears to be mediated by a different mechanism which involves the renal nerves. This is because centrally administered hypertonic saline depresses renal nerve activity as well as renin secretion in sheep, the effect on renin secretion is blocked by renal denervation, and both of these response measures are abolished by ablation of the lamina terminalis (May et al. 2000; McKinley et al. 2001).

One consideration that has been difficult to resolve is whether osmoreceptor neurons in the lamina terminalis which have axonal connections to the supraoptic and paraventricular nuclei, and presumably subserve vasopressin and oxytocin secretion in the rat, are separate from those which may influence other functions such as thirst, renal sodium excretion and renin secretion. Evidence which favours the idea of a common set of neurons is the observation that some neurons within the OVLT that project to the supraoptic nucleus have collateral axons that branch to other brain regions (Oldfield et al. 1992).

Although the discussion to this point has focused on the osmoreceptor role of the OVLT and subfornical organ; there is evidence of a possible osmoregulatory role of the area postrema in the dog and rat. Dogs in which the area postrema had been ablated did not excrete a hypertonic saline load as rapidly as normal dogs, although they excreted water normally in response to a water load and concentrated urine normally in response to dehydration (Wise and Ganong 1960). Rats with ablation of the area postrema also did not excrete a hypertonic saline load normally (Curtis et al. 1999), and although rats have an increased water intake after 24 h of water deprivation, this is insufficient to compensate for a greater negative water balance because a larger volume of urine is lost in these animals during this time as a result of the excretion of a more dilute urine (Hyde and Miselis 1984). In the rat, there is also increased expression of *c-fos* in neurons of the area postrema in response to intravenous infusion of hypertonic saline (Hochstenbach and Ciriello 1986; Rinaman et al. 1997). Although some reservations must be held that the increase in Fos expression may have been the result of arterial pressure increasing with intravenous hypertonic saline, increased Fos in the area postrema also occurred in dehydrated rats, which have hypertonic plasma but unchanged arterial pressure (McKinley and Oldfield, unpublished observations). There is also evidence that rats in which the area postrema is ablated do not regulate oxytocin and vasopressin secretion in response to increased plasma sodium concentration (Huang et al. 2000). Additionally, ablation of the area postrema in rats has been shown to disrupt osmotically stimulated drinking (Curtis et al. 1999).

8.1.3
Influences of Relaxin on Osmoregulation

Binding sites for the hormone relaxin have been identified in both the subfornical organ and the OVLT of the rat, as well as in other brain regions such as the neocortex, dorsal tegmentum and supraoptic and paraventricular nuclei (Osheroff and Phillips 1991; Tan et al. 1999). This peptide hormone, which is secreted from the ovary during pregnancy, was first shown to stimulate vasopressin secretion and water drinking and to cause a pressor response when injected into the lateral ventricles of rats (Parry et al. 1990, 1994; Thornton and Fitzsimons 1985; Summerlee and Robertson 1985). When administered intravenously, relaxin also stimulates vasopressin secretion (Geddes et al. 1994) and water drinking in rats (Sinnayah et al. 1999). These effects can be inhibited by centrally administered angiotensin AT_1 receptor antagonists (Parry et al. 1991; Geddes et al. 1994). Longer-term peripheral administration of relaxin causes a sustained reduction in plasma osmolality in ovariectomised rats (Weisinger et al. 1993), and it has been suggested that there is resetting of the central osmostat to a lower plasma osmolality (Lindheimer et al. 1989). In view of the presence of relaxin binding sites in the subfornical organ and OVLT, it seems likely that its dipsogenic and vasopressin-secreting actions are mediated by these sensory CVOs.

Intravenously infused relaxin causes a marked increase in the number of neurons expressing Fos in the outer shell of the subfornical organ and dorsal cap of the OVLT (Fig. 9). It has been shown recently that many of these Fos-labelled neurons also project to the supraoptic and paraventricular nuclei of the hypothalamus and are activated by intravenously infused relaxin. This evidence suggests that relaxin acts on these two CVOs to influence vasopressin secretion (Sunn et al. 2001). We have observed electrophysiologically that relaxin directly stimulates neurons in slices of the subfornical organ in vitro (Sunn et al. 2002). In addition, we have shown that relaxin also acts on the subfornical organ to stimulate thirst, because ablation of the subfornical organ impairs relaxin-induced drinking (Sunn et al. 2002). Such ablation also prevents the pressor response to centrally administered relaxin (Mumford et al. 1989).

8.2 Central Actions of Circulating Angiotensin

8.2.1 Water Drinking

The first experimental evidence indicating a role for forebrain sensory CVOs in regulating fluid balance was the finding that ablation of the subfornical organ abolished angiotensin-induced drinking in the rat. After the pioneering studies of James Fitzsimons and his colleagues, who showed that increases in the circulating levels of angiotensin II could induce water drinking in rats, the question arose as to how an octapeptide like angiotensin II could stimulate the brain if it did not penetrate the blood-brain barrier. John Simpson and his colleagues proposed that it might act at a circumventricular organ. They went on to show that the subfornical organ was sensitive to picomoles of angiotensin II injected into it and that ablation of the subfornical organ of the rat largely prevented the drinking in response either to intravenous infusions of angiotensin II or to conditions which caused circulating levels of angiotensin II to increase as a consequence of renin secretion being stimulated. Such conditions included isoproterenol administration and hypovolaemia resulting from subcutaneous administration of colloid (polyethylene glycol) (Simpson and Routtenberg 1973; Simpson et al. 1978).

Electrophysiological recordings from the subfornical organs of anaesthetised cats and rats showed that angiotensin II directly stimulated neurons in this CVO (Felix 1976; Phillips and Felix 1976; Gutman et al. 1988). Later, in vitro autoradiographic studies showed that the subfornical organ was rich in angiotensin binding sites (Mendelsohn et al. 1984; Speth et al. 1985; McKinley et al. 1986a, 1987a). More recently, competitive binding studies, in situ hybridisation histochemistry, and immunohistochemistry (Bunnemann et al. 1992; Lenkei et al. 1997; Giles et al. 1999) have identified those binding sites as receptors of the AT_1 subgroup. As expected, administration of selective AT_1 antagonists such as losartan, valsartan or ZD7155 blocked angiotensin II-induced drinking, electrophysiological responses and *c-fos* expression (Dourish et al. 1992; Beresford and Fitzsimons 1992; Oldfield and McKinley 1994; McKinley et al. 1996b; Riediger et al. 1999).

A seemingly contradictory result was the observation that ablation of the subfornical organ of the rat had little influence on drinking in response to infusion of angiotensin II into the cerebral ventricles but ablation of the AV3V region (which contains the OVLT) abolished this response (Buggy and Johnson 1979). In addition, when angiotensin II infused into the lateral ventricle was prevented from reaching the anteroventral chamber of the third ventricle, once again water drinking did not occur (Buggy and Fisher 1977). The resolution to this paradox came when it was realised that intracerebroventricular angiotensin II acted largely at sites other than the subfornical organ, in particular the median preoptic nucleus, a site also rich in angiotensin AT_1 receptors (Allen et al. 2000). This can be seen particularly well when the pattern of Fos expression in response to ICV angiotensin II is compared to that obtained with intravenous infusion of angiotensin II (McKinley et al. 1995). Although intravenous angiotensin II causes extensive activation of neurons in the subfornical organ and OVLT, very few neurons increase Fos expression in the MnPO. By contrast,

Fos expression increases strongly in the MnPO and OVLT after ICV angiotensin II, but only in the ventral periventricular parts of the subfornical organ (McKinley et al. 1995). Both electrolytic or chemical ablation (which leave fibre tracts intact) of the MnPO and cutting fibre pathways between the subfornical organ and the MnPO cause angiotensin-induced drinking to be abolished (Mangiapane et al. 1983; Cunningham et al. 1992).

These data suggest that the neural pathway from the angiotensin-sensitive neurons of the subfornical organ to the MnPO is part of the neural circuitry underpinning angiotensin-induced drinking. In this regard, the neurons of the subfornical organ that project to the MnPO are predominantly in the outer shell of the subfornical organ, suggesting that this group of neurons, not those in the core, mediate angiotensin-induced drinking.

Up to this point, the data discussed have been overwhelmingly obtained from studies on laboratory rats. However, evidence from studies of injections of angiotensin II into the subfornical organ and preoptic region of dogs (Fitzsimons and Kucharczyk 1978) indicates the likelihood that angiotensin receptors in the subfornical organ also influence water drinking in dogs. In sheep it was shown that neither ablation of the subfornical organ nor OVLT inhibited drinking response to systemic or intracerebroventricular infusion of angiotensin II (McKinley et al. 1982, 1986d). However, ablation of both CVOs with parts of the MnPO did prevent such responses (McKinley et al. 1986c). These findings suggest that both OVLT and the subfornical organ have a dipsogenic role in sheep but that there is considerable redundancy of function between these CVOs in this species.

8.2.2
Angiotensin-Induced Vasopressin Secretion

Systemic infusion of angiotensin II may also stimulate the secretion of vasopressin if high enough blood levels are achieved (Ramsay et al. 1978; Reid 1984). Ablation of the subfornical organ in rats, or of the OVLT in dogs, severely disrupts this response (Mangiapane et al. 1984; Thrasher 1985). Electrophysiological experiments show that a pathway from the subfornical organ via a synapse in the MnPO may signal to the paraventricular nucleus (and probably supraoptic nucleus) to increase the secretion of vasopressin (Tanaka et al. 1987). However, there is also evidence that direct pathways from the subfornical organ to supraoptic and paraventricular nuclei may be activated by systemic angiotensin II (Ferguson and Renaud 1987). Some neurons in the peripheral shell of the subfornical organ and in the dorsal cap of the OVLT, which project to the supraoptic nucleus, increase the expression of *c-fos* in response to intravenous angiotensin II. Neurons in the OVLT and SFO which have direct axonal input to vasopressin-secreting cells of the supraoptic nucleus (Oldfield et al. 1994) may thus mediate the vasopressin response to intravenous angiotensin II. A diagram of the neural pathways which may mediate effects of circulating angiotensin II on CVOs to influence water and sodium balance is shown in Fig. 27.

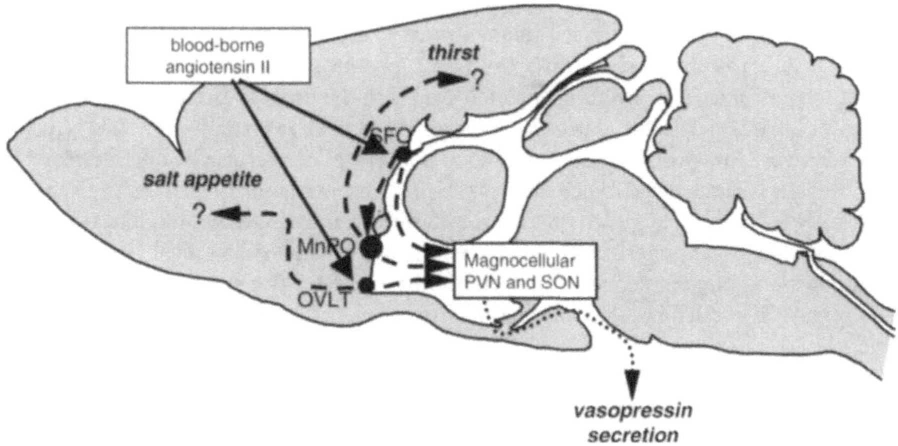

Fig. 27 Diagram of neural pathways in the rat brain that may mediate thirst, sodium appetite and vasopressin secretion following stimulation by circulating angiotensin II of neurons in the circumventricular organs of the lamina terminalis. The cortical regions mediating thirst and sodium appetite are unknown, and this is indicated by the *question marks*. *MnPO*, Median preoptic nucleus; *OVLT*, organum vasculosum of the lamina terminalis; *PVN*, hypothalamic paraventricular nucleus; *SFO*, subfornical organ; *SON*, supraoptic nucleus

8.2.3
Sodium Hunger

For some years now, evidence has been accumulating that circulating angiotensin II can stimulate sodium hunger in rats and sheep (Findlay and Epstein 1980; Weisinger et al. 1987; Thunhorst and Fitts 1994; Denton et al. 1996). For example, in both of these species, the sodium appetite resulting from sodium depletion is inhibited by sufficiently high doses of peripherally administered captopril, an ACE inhibitor which reduces circulating levels of angiotensin II. However, sodium appetite is restored if an intravenous infusion of angiotensin II is administered (Weisinger et al. 1987, 1996). Because circulating angiotensin II appeared to have a role in the genesis of salt hunger, we postulated that it acted on a sensory CVO to achieve this response (Weisinger et al. 1990). The feasibility of this proposal was shown by evidence that ablation of the subfornical organ depressed the sodium intake in response to diuretic-induced loss of salt in rats (Weisinger et al. 1990; Thunhorst et al. 1999). However, although microinjections of angiotensin II into the subfornical organ stimulate thirst, they do not cause a salt appetite, which does not support the earlier proposal of the subfornical organ mediating angiotensin II-induced salt appetite (Fitts et al. 2000). It should be noted, though, that injection of angiotensin II into the subfornical organ of rats does stimulate a salt hunger if serotonin receptors in the lateral parabrachial nucleus are blocked. This result suggests that there may be inhibitory signals from the brain stem that depress salt appetite mediated by angiotensin action on the subfornical organ (Columbari et al. 1996).

The OVLT has also been implicated in the generation of salt appetite. Ablation of the OVLT region in rats depresses sodium appetite in response to sodium depletion

(Chiaraviglio and Perez Guaita 1984), whereas local injection of angiotensin II into the OVLT region stimulates NaCl intake in rats (Fitts and Masson 1990). In addition, injection of captopril directly into the OVLT prevents the enhancement of sodium appetite that occurs in sodium-depleted rats that are treated peripherally with low doses of an ACE inhibitor (Fitts and Masson 1990). The explanation for this effect is that relatively low doses of captopril block peripheral conversion of angiotensin I to angiotensin II but do not block the very high concentrations of ACE in the sensory CVOs. High levels of angiotensin I then accumulate in the circulation, and these can be converted to angiotensin II locally in the OVLT and subfornical organ, which stimulates angiotensin-sensitive neurons in these CVOs. The extensive expression of Fos seen in the sensory CVOs after low-dose captopril treatment is consistent with this idea (McKinley et al. 1997b). When high concentrations of ACE inhibitors are injected directly into the CVOs, however, the local production of angiotensin II there is inhibited. The data suggest that local production of angiotensin II in the OVLT stimulates salt hunger whereas that in the subfornical organ mainly stimulates thirst (Fitts and Masson 1990). In support of this idea, we found that ablation of the subfornical organ did not prevent the enhanced sodium appetite following low-dose captopril treatment (Weisinger et al. 1990). Interestingly, although Menani et al. (1998) found that injection of the AT_1 antagonist losartan into the subfornical organ did not reduce the sodium appetite due to combined diuretic and peripheral low-dose captopril treatment, it did prevent the further increase in sodium intake that occurs after this treatment in rats which also had been injected with the serotonin antagonist methysergide in the lateral parabrachial nucleus. These data suggest that an action of angiotensin II on the subfornical organ may have a role in sodium intake of rats when inhibitory influences from the brain stem are removed.

Rats in which the area postrema has been ablated develop a polydipsia and overconsumption of hypertonic saline solution, and it was suggested originally that this was a response to the sodium lost in urine as a consequence of ablation of the area postrema (Hyde and Miselis 1984). However, it has been suggested that the large saline intake of rats with area postrema lesions is the result of disruption of inhibitory signals from this CVO (Curtis et al. 1999; Stricker et al. 2001). Another aspect of possible involvement of the CVOs in sodium hunger is a recent report that a voltage-gated sodium channel (NaG/Na_v2) which is expressed in the subfornical organ and OVLT may have a role in the regulation of salt intake. If the gene for this channel is 'knocked out' in mice, they have an exaggerated hunger for salt in response to dehydration or sodium depletion. This is concomitant with overactivity of neurons in the subfornical organ and OVLT (Watanabe et al. 2000).

8.3 Influences on Arterial Pressure

8.3.1 Area Postrema

Although Wise et al (1960) reported that ablation of the area postrema in dogs tended to result in a lower arterial pressure after haemorrhage, the main work which stimulated interest in the area postrema as a cardiovascular regulatory region of the brain was the observation that infusions of angiotensin II into the vertebral artery of anaesthetised dogs increased arterial pressure more effectively than intravenous infusion (Bickerton and Buckley 1961; Ferrario et al. 1970; Joy and Lowe 1970). This also occurred in conscious dogs (Reid 1984), and because the area postrema is the only sensory CVO supplied by the vertebral artery in the dog, it was the prime candidate as the central site of action of angiotensin II infused into this blood vessel. Indeed, ablation of the area postrema was shown to abolish the pressor effects of intervertebral artery infusion of angiotensin II in the dog (Joy and Lowe 1970; Ferrario et al. 1972). Pressor responses to intravenous infusion of angiotensin II are also reduced by ablation of the area postrema in dog (Ferrario et al. 1979). In rats, ablation of the area postrema did not affect the pressor response to acute systemic infusion of angiotensin II (Haywood et al. 1980; Fink et al. 1987), and it may be relevant that the density of angiotensin binding sites appears to be relatively greater in the dog (Speth et al. 1985) than in a number of other species. However, ablation of the area postrema does depress pressor responses to chronically infused angiotensin II in rats (Fink et al. 1987), as it does in dogs (Otsuka et al. 1986), and both renovascular hypertension and genetic spontaneous hypertension are attenuated by ablation of the area postrema in the rat (Fink et al. 1986; Rittenhouse et al. 1986).

The neural connectivity of the area postrema, as reviewed in Sect. 6, shows this CVO to have direct efferent connections to the NTS and RVLM, brain sites that play crucial roles in the sympathetic control of the cardiovascular system. It also brings into focus a frequently mentioned criticism of experiments employing ablation of the area postrema. Because the area postrema is adjacent to the NTS, how much of the effect of area postrema ablation is due to damage to the adjacent NTS? Of interest in this regard are observations that direct injections of angiotensin II into either the area postrema or the immediately adjacent medial subnucleus of the NTS both increase arterial pressure and heart rate (Averill et al. 1985). It should be noted that dendritic processes coming from neurons within the medial NTS (a site rich in angiotensin receptors) are thought to extend into the area postrema. Perhaps the important factor in these considerations is the absence of the blood-brain barrier in the area postrema but not the NTS. Blood-borne angiotensin II has been shown to directly stimulate a group of neurons in the area postrema independently of blood pressure changes (Papas et al. 1990). In addition, neurons within the medial NTS with dendrites extending into the area postrema may also be stimulated by circulating angiotensin II. Electrical stimulation of the area postrema may either increase or decrease arterial pressure, depending on the frequency of stimulation (Ferguson and Marcus 1988; Hasser et al. 1987). The effect of low-frequency stimulation was different to the effect obtained with stimulation of the medial NTS at a similar frequency, suggesting that different neural elements were involved.

Early studies indicated that angiotensin's area postrema-mediated pressor action was due to increased sympathetic nervous activity (Ferrario et al. 1987). More recent work has refined this to include the idea that chronically administered angiotensin II acts on the area postrema to reset the arterial baroreceptor reflex to a higher level of mean arterial pressure so that sympathetic activity remains elevated at high blood pressures (Cox and Bishop 1991; Bishop and Sanderford 2000). As an example for this, it has been shown in rats that during the first 2 h of intravenous infusion of angiotensin II there is increased blood pressure and activation of caudal ventrolateral medulla (CVLM) neurons (as indicated by Fos) which mediate the baroreflex but not neurons in the RVLM (Hasser et al. 2000). These authors also showed that after 18 h of intravenously administered angiotensin II, neurons in CVLM were no longer activated, indicating a resetting of the baroreceptor reflex. However, by this time, neurons in the RVLM were activated, and this response did not occur if the area postrema had been ablated and the arterial pressure did not remain elevated (Hasser et al. 2000). Because sympathetic premotor neurons are most abundant in the RVLM, it is likely that chronically infused angiotensin II results in their long-term activation, even though arterial pressure was high.

Vasopressin secreted into the circulation by the posterior pituitary may also influence cardiovascular function via an action on the area postrema. Undesser et al. (1985) produced evidence that vasopressin acts on the area postrema to potentiate the baroreceptor reflex. Because of this effect, vasopressin, although having a potent direct vasoconstrictor action on vascular smooth muscle, does not raise arterial pressure to the extent that may be expected for such a vasoconstrictor agent.

8.3.2 OVLT

The OVLT is a central component of the 'AV3V' region, which has been implicated for many years now in the genesis of various forms of experimental hypertension (Brody et al. 1978) and in the pressor response to centrally administered angiotensin II (Johnson et al. 1978). However, the AV3V has several other component nuclei (e.g. median preoptic nucleus) which could mediate these effects in addition to the OVLT and which also exhibit angiotensin II AT_1 receptors. Therefore, the role of the OVLT in AV3V-mediated effects on cardiovascular variables is unclear.

A more precise series of neural stimulations in urethane-anaesthetised rats by Mangiapane and Brody (1987) showed that with electrical stimulation of sites within the AV3V, the OVLT was the site from which pressor and vasoconstrictor responses could be obtained; at adjacent regions such as the ventral median preoptic, anterior median preoptic or periventricular preoptic nuclei, depressor and vasodilator responses were obtained. The observation that ganglionic blockade prevented increased arterial pressure and increased vascular resistance in the mesenteric and renal vascular beds in response to electrical stimulation of the OVLT suggested that neurons in the OVLT influence sympathetic vasoconstrictor pathways (Mangiapane and Brody 1987).

Recently, it was shown that direct injection of NO donors into the OVLT region of anaesthetised rats caused a large reduction in arterial pressure (as much as 55 mmHg). Inhibition of NO synthase by injection of N^G-nitro-L-arginine methyl ester (L-NAME) into the OVLT increased arterial pressure. These effects were abolished by spinal transection (Lin et al. 1999). These data suggest that the OVLT can influence the cardiovascular system by altering sympathetic nervous activity.

8.3.3 Subfornical Organ

Stimulation of the subfornical organ either electrically (Mangiapane 1987; Ferguson and Renaud 1985) or by microinjection of angiotensin II or carbachol increases arterial pressure (Mangiapane and Simpson 1980, 1983). In addition, ablation of the subfornical organ reduces the pressor response to acute systemic infusion of angiotensin (Mangiapane and Simpson 1980).

The efferent neural circuitry that could mediate such pressor responses may involve a pathway to the hindbrain or spinal cord via the hypothalamic paraventricular nucleus (Ferguson and Renaud 1987); ablation of this nucleus severely depresses the pressor response to electrical stimulation of the subfornical organ (Ferguson and Renaud 1984).

8.3.4 The Extended Autonomic Neuraxis

Miselis et al. (1987) outlined the way in which the sensory circumventricular organs are integrated into a visceral neuraxis, distributed from the hindbrain to the forebrain, enabling the CVOs to influence endocrine, autonomic and ingestive function. They proposed that interoreceptive inputs from the CVOs were integrated with visceral sensory inputs from the periphery, within this neuraxis, to modulate its motor outputs. Retrograde trans-synaptic tracing of neural pathways back into the CNS from various peripheral sites with a neurotropic virus shows that there are polysynaptic pathways from the forebrain sensory CVOs to several organs and tissues (Fig. 23). Injection of pseudorabies virus into peripheral sympathetic ganglia also results in the eventual retrograde labelling of the subfornical organ and OVLT, as well as other forebrain sites (Westerhaus and Leowy 1999). Recently, we showed that there are a few neurons in the subfornical organ, mainly in its outer shell, and the OVLT, mainly in its lateral zone, which were both polysynaptically labelled from the kidney and expressed the angiotensin AT_1 receptor (Giles et al. 2001). This suggests that circulating angiotensin II could influence the sympathetic nerve supply to the kidney via influences on these two CVOs. We propose that there are polysynaptic connections from neurons in the outer shell region of the subfornical organ, and the lateral zone of the OVLT, to peripheral sympathetic vasomotor neurons, and these may influence cardiovascular function. Together with other forebrain sites (e.g. the median preoptic nucleus, bed nucleus of the stria terminalis), the CVOs may interact with hypothalamic, midbrain and hindbrain nuclei to participate in an extended autonomic neuraxis (Miselis et al. 1987; Sly et al. 1999; Westerhaus and Leowy 1999). A diagram of the neural pathways that may mediate the actions of circulating angiotensin II on CVOs to influence cardiovascular function is given in Fig. 28.

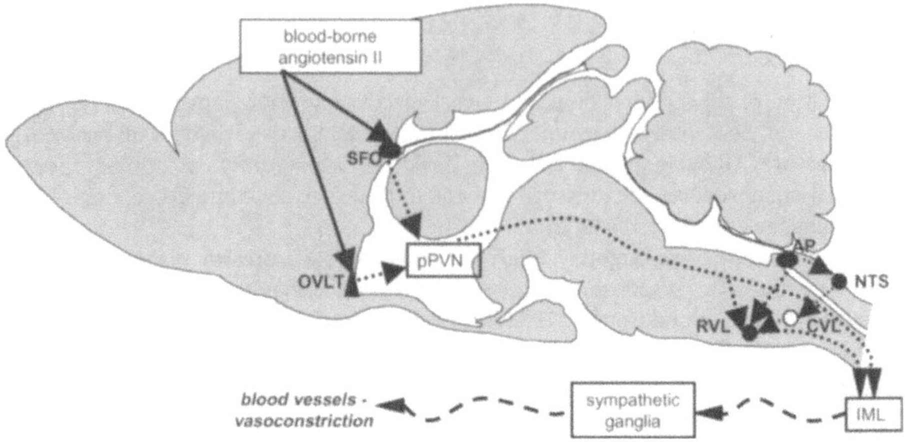

Fig. 28 Diagram of neural pathways in the rat brain that may influence cardiovascular function following stimulation of neurons in the subfornical organ, OVLT and area postrema by blood-borne angiotensin II. The pathway from CVL to RVL indicated by the *black arrow* is inhibitory. *AP*, area postrema; *CVL*, caudal ventrolateral medulla; *IML*, intermediolateral cell column of the spinal cord; *NTS*, nucleus of the solitary tract; *OVLT*, organum vasculosum of the lamina terminalis; *pPVN*; parvocellular region of the hypothalamic paraventricular nucleus; *RVL*, rostral ventrolateral medulla; *SFO*, subfornical organ; *SON*, supraoptic nucleus

8.4
Area Postrema and Vomiting

The proposal that the area postrema is a chemoreceptor 'trigger zone' (Borrison and Brizzee 1951) for the vomiting reflex was one of the first functions to be attributed to a sensory CVO. The initial evidence supporting this idea came from studies in cats which showed that ablation of the area postrema was an effective way of preventing vomiting following the administration of emetic agents. This has been confirmed many times with a variety of emetic agents in several species, including humans. However, ablation of the area postrema does not prevent some emetic stimuli from being effective e.g. motion sickness and vagal stimulation (see Miller and Leslie 1994 for review). In regard to humans, cancer patients who were receiving chemotherapy and experiencing intractable nausea, vomiting, retching or hiccups underwent surgery for the thermal ablation of the area postrema; this resulted in a remarkable extinction of the nausea, vomiting and hiccups, although several of the patients died subsequently as a result of the malignancies. Post-mortem examination revealed that the area postrema had been destroyed but additionally some part of the adjacent NTS was damaged (Lindstrom and Brizzee 1962). This destruction of the immediately adjacent NTS is almost always present with lesions of the area postrema and makes precise evaluation of the role of the area postrema difficult.

Recent studies utilising the expression of *c-fos* as a marker of neuronal activity show that systemic administration of stimuli such as cisplatin and also vagal stimulation do result in activation of neurons in the area postrema as well as the NTS and medullary pre-motor and motor pathways for vomiting (see Sect. 7.8). The complex series of co-ordinated muscle contractions that are necessary for vomiting has led to the consideration, that once triggered, a medullary pattern generator takes over the co-ordination of this activity (Carpenter 1990). The vagal afferents terminating in the NTS are essential for vomiting, and serotonin 5-HT_3 antagonists that effectively suppress nausea and vomiting in cancer patients act at the distal end of these vagal afferents (Andrews et al. 1990; Blessing 1997). The precise role of the area postrema for vomiting still remains to be determined, but if it is a sensor for circulating agents because it lacks a blood-brain barrier, the neural connection to the adjacent medial subnucleus of the NTS is probably crucial.

8.5 Neuroimmune Responses and Fever

All three sensory CVOs have been implicated in the production of fever and other CNS-mediated aspects of the acute-phase response to systemic infection or to circulating pyrogen. Initially, it was pointed out by Blatteis et al. (1983) that the blood-brain barrier would prevent the passage of large pyrogenic molecules such as lipopolysaccharides and interleukins, exogenous and endogenous pyrogens, respectively, into the brain. They proposed that a CVO, lacking the blood–brain barrier, may be the site at which pyrogens act to induce fever. They went on to show in guinea pigs that ablation of the AV3V region, which included the OVLT, prevented the febrile response to peripherally administered lipopolysaccharide. This lesion did not affect the pyrogenic response to endogenous pyrogen injected into the preoptic region; however, it was also shown in sheep that ablation of tissue in the anterior wall of the third ventricle, including the OVLT, prevented febrile responses (Blatteis et al. 1987). These lesions were relatively large and included considerably more periventricular tissue than just the OVLT. In sheep, we have made much smaller lesions confined to the OVLT and still observed robust febrile responses in these animals (G.L. Pennington and M.J. McKinley, unpublished observations). By contrast, studies in the rat and rabbit (Stitt 1985) showed that ablation of the OVLT enhanced febrile responses to various pyrogens (lipopolysaccharide, zymosan, endogenous pyrogen). In the rabbit, however, ablation of the OVLT reduced but did not abolish febrile responses to intravenous injection of interleukin (IL)-1β or tumour necrosis factor (TNF)-α, whereas it prolonged the fever in response to intracerebral prostaglandins. (Hashimoto et al. 1994). Stitt and Yamada (1989) showed in rats that if microinjections of relatively minute amounts (less than 1/300–1,000 the effective intravenous dose) of the pyrogens were given into the OVLT 3 days earlier, a pronounced enhancement of febrile responses to systemically administered endogenous pyrogen occurred. The potentiating effect was not observed when the same low dose of pyrogen was injected into the adjacent medial preoptic region 3 days earlier. These authors proposed that endogenous pyrogens, released from macrophages, enter the OVLT to stimulate prostaglandins release; these prostaglandins then diffuse into the adjacent neuropil of the thermoregulatory medial preoptic area to influence later pyrogenic responses.

Later studies in which the OVLT was discretely ablated in the rat (Takahashi et al. 1997) were unable to show any change in the febrile response to intravenous lipopolysaccharide, whereas ablation of the another sensory CVO, the subfornical organ, did result in change. Ablation of the area postrema also was ineffective at suppressing the febrile response in this study (Takahashi et al. 1997). Lesions of the area postrema cause a marked suppression of IL-1-induced ACTH release (Lee et al. 1998), as does ablation of the OVLT (Katsurra et al. 1990). Recently, it has been shown that injection of tyrosine kinase inhibitors into the OVLT inhibited the increased iNOS expression and fever that had been stimulated by injection of lipopolysaccharide into the OVLT (Lin and Lin 2000).

Generation of prostaglandin (PG) in the brain is an essential step in generating a fever. Concerning the possibility that PG could act on neurons in the OVLT as well as the medial preoptic area, electrophysiological studies show that neurons within, or very near, the OVLT are directly stimulated by PGE_2 (Matsuda et al. 1992; Hori et al. 1992; Hori et al. 2001). Injection of PGE_1 into the OVLT can produce fever, and this CVO is extremely sensitive to PGE_1 in this regard (Stitt 1986; Stitt 1991). PGE_2 binding sites are observed in the OVLT region (Matsumura et al. 1990; Matsumara and Kobayashi 2001), and various subtypes of PGE_2 receptor (EP_1, EP_2, EP_3 and EP_4 receptors) have been identified in many brain regions. The EP_3 receptor has been identified by immunohistochemistry in the OVLT (Nakamura et al. 1999) and by in situ hybridisation in the subfornical organ but not in the area postrema (Ek et al. 2000). In regard to the EP_2 receptor, exogenous pyrogen administration resulted in the expression of its mRNA (detected by in situ hybridisation) increasing substantially in the OVLT, area postrema and subfornical organ, as well as in several other brain regions (Zhang and Rivest 1999). Of the sensory CVOs, only the subfornical organ appeared to increase expression of EP_4 mRNA in response to pyrogens (Zhang and Rivest 1999).

The expression of the transcription factors *c-fos* and K$B\alpha$ in response to systemic pyrogens is also markedly increased in all three sensory CVOs as well as several other brain regions (Sagar et al. 1995; Quan et al. 1997). A number of other key molecules that participate in the cascade that follows pyrogen administration have also been shown to be generated within the three sensory CVOs. These molecules include the cytokines IL-1β in the OVLT (Nakamori et al. 1993) and other CVOs (Quan et al. 1998); TNF-α (Breder et al. 1994; Nadeau and Rivest 1999; Quan et al. 1999; Jhang and Rivest 2001) and IL-6 (Valleries and Rivest 1997). In addition, mRNA for IL receptors IL-IR type 1, IL-6Rα and gpL 130 are expressed in CVOs (Cao et al.1996; Vailleries and Rivest 1997) in response to pyrogen.

Other key molecules linking this cascade are the cell membrane receptor CD-14, which binds the lipopolysaccharide-LPS binding protein complex, and Toll-like receptor 4 (TLR-4), with which it may interact for signal transduction. Although mRNAs encoding both of these molecules are expressed under control conditions in all CVOs and in the meninges (Lacroix et al. 1998; Laflamme and Rivest 2001), CD-14 mRNA expression greatly increases in the CVOs with lipopolysaccharide treatment (Lacroix et al. 1998) whereas TLR-4 is inhibited (Laflamme and Rivest 2001). Within 3 h after pyrogen treatment, there is also a marked increase in the expression of the suppressor of cytokine signalling-3 (SOCS-3) molecule, particularly in the subfornical organ, OVLT, choroid plexus, area postrema, median eminence, arcuate nucleus and hypothalamic paraventricular nucleus. The above areas are all richly vascularised, and, apart from the paraventricular nucleus, their capillaries also exhibit a fenestrated endothelium. By 6 h, this expression of SOCS-3 occurs in many blood vessels throughout the brain (Lebel et al. 2000).

Thus it is clear that the CVOs, as well as meninges and choroid plexus, are significant sites in the brain for synthesis and utilisation of the molecular cascades that subserve the febrile response. Perhaps this reflects the enormous number of blood vessels and the diminished blood-brain barrier in these sites, because the capillary endothelium is the site of many of the crucial reactions. The question that now remains is whether these cytokine-transcription factor-prostaglandins cascades within the CVOs, which result in PGE_2 production there, permit diffusion of PGE_2 to adjacent thermoregulatory brain sites, or whether they activate specific neural groups in the sensory CVOs to transmit pyrogenic signals to other brain regions.

8.6 Concluding Remarks

Despite the small amount of space that the sensory CVOs occupy in the central nervous system, there seems little doubt that they fulfil functions that contribute considerably to survival. They have important roles in controlling the intake and excretion of salt and water and the expulsion of deleterious substances. They may also participate in regulation of cardiovascular function and some neuroimmune responses. All three sensory CVOs may participate in regulating fluid and electrolyte homeostasis, cardiovascular control and the febrile response. It remains to be determined how much redundancy of function there is across these three CVOs, particularly as there are common sites in the brain that they all link with.

Several aspects of the sensory CVOs still remain to be elucidated. The functions of most of the various receptors that have been detected in the CVOs, and the role of many of the neural pathways connecting them with the rest of the brain, remain unknown. The sensory CVOs should remain fertile ground for future research.

References

Akert K, Pfenninger K, Sandri C (1967) The fine structure of the synapses in the subfornical organ of the cat. Z. Zellforsch 81:537–556

Akert K, Potter HD, Anderson JW (1961) The subfornical organ in mammals. J. Comp. Neurol. 116:1–14

Allen AM, Chai SY, Clevers J, McKinley MJ, Paxinos G, Mendelsohn FA (1988) Localization and characterization of angiotensin II receptor binding and angiotensin converting enzyme in the human medulla oblongata. J Comp Neurol 269:249–64

Allen AM, McKinley MJ, Mendelsohn FA (1988) Comparative neuroanatomy of angiotensin II receptor localization in the mammalian hypothalamus. Clin Exp Pharmacol Physiol 15:137–145

Allen AM, Oldfield BJ, Giles ME, Paxinos G, McKinley MJ, Mendelsohn FAO (2000) Localization of angiotensin receptors in the nervous system. In: Quirion R, Björklund A, Hökfelt T (eds) Peptide Receptors, Part 1. Elsevier Science, pp 79–124 vol 16

Alm P, Skagerberg G, Nylen A, Larsson B, Andersson KE (1997) Nitric oxide synthase and vasopressin in rat circumventricular organs. An immunohistochemical study. Exp Brain Res 117:59–66

Anderson WA, Bruni JE, Kaufmann A (1990) Afferent connections of the rat's supraoptic nucleus. Brain Res Bull 24:191–200

Andersson B, Leksell LG, Lishajko F (1975) Perturbations in fluid balance induced by medially placed forebrain lesions. Brain Res 99:261–275

Andres KH (1965) Der Feinbau des Subfornikalorganes vom Hund. Z Zellforsch 68:445–473

Andrews PL, Davis CJ, Bingham S, Davidson HI, Hawthorn J, Maskell L (1990) The abdominal visceral innervation and the emetic reflex: pathways, pharmacology, and plasticity. Can J Physiol Pharmacol 68:325–345

Anthes N, Schmid HA, Hashimoto M, Riediger T, Simon E (1997) Heterogeneous actions of vasopressin on ANG II-sensitive neurons in the subfornical organ of rats. Am J Physiol 273:R2105–2111

Antunes-Rodrigues J, McCann SM, Rogers LC, Samson WK (1985) Atrial natriuretic factor inhibits dehydration- and angiotensin II- induced water intake in the conscious, unrestrained rat. Proc Natl Acad Sci U S A 82:8720–8723

Ariumi H, Saito R, Nago S, Hyakusoku M, Takano Y, Kamiya H (2000) The role of tachykinin NK-1 receptors in the area postrema of ferrets in emesis. Neurosci Lett 286:123–126

Armstrong DM, Miller RJ, Beaudet A, Pickel VM (1984) Enkephalin-like immunoreactivity in rat area postrema: ultrastructural localization and coexistence with serotonin. Brain Res 310:269–278

Armstrong DM, Pickel VM, Joh TH, Reis DJ (1982a) Electron microscopic immunocytochemical localization of tyrosine hydroxylase in the area postrema of rat. J Comp Neurol 206:259–272

Armstrong DM, Pickel VM, Reis DJ (1982b) Electron microscopic immunocytochemical localization of substance P in the area postrema of rat. Brain Res 243:141–146

Atkinson L, Batten TF, Deuchars J (2000) P2X(2) receptor immunoreactivity in the dorsal vagal complex and area postrema of the rat. Neuroscience 99:683–696

Averill DB, Diz DI, Barnes KL, Ferrario CM (1985) Pressor and tachycardic actions of angiotensin II in the nucleus tractus solitarii (NTS) region of the dog. Fed Proc 44:1554

Badoer E, McKinley MJ, Oldfield BJ, McAllen RM (1992) Distribution of hypothalamic, medullary and lamina terminalis neurons expressing Fos after hemorrhage in conscious rats. Brain Res 582:323–328

Badoer E, McKinley MJ, Oldfield BJ, McAllen RM (1993a) A comparison of hypotensive and non-hypotensive hemorrhage on Fos expression in spinally projecting neurons of the paraventricular nucleus and rostral ventrolateral medulla. Brain Res 610:216–223

Badoer E, McKinley MJ, Oldfield BJ, McAllen RM (1994) Localization of barosensitive neurons in the caudal ventrolateral medulla which project to the rostral ventrolateral medulla. Brain Res 657:258–268

Badoer E, Oldfield BJ, McKinley MJ (1993b) Haemorrhage-induced production of Fos in neurons of the lamina terminalis: role of endogenous angiotensin II. Neurosci Lett 159:151–154

Barry J, Hoffman GE, Wray S (1985) LHRH-containing systems. In: Bjorklund A, Hokfelt T (eds) Handbook of Chemical Neuroanatomy. Vol. 4. GABA and Neuropeptides in the CNS. Part I. Elsevier, Amsterdam, 166–217 pp

Barth SW, Gerstberger R (1999) Differential regulation of angiotensinogen and AT1A receptor mRNA within the rat subfornical organ during dehydration. Brain Res Mol Brain Res 64:151–64

Bayer SA, Altman J (1987) Development of the preoptic area: time and site of origin, migratory routes, and settling patterns of its neurons. J Comp Neurol 265:65–95

Behnsen G (1927) Über die Farbstoffspeicherung im Zentralnervensystem der weiβen Maus in verschiedenen Alterszuständen. Z Zellforsch 4:515–572

Beresford MJ, Fitzsimons JT (1992) Intracerebroventricular angiotensin II-induced thirst and sodium appetite in rat are blocked by the AT1 receptor antagonist, Losartan (DuP 753), but not by the AT2 antagonist, CGP 42112B. Exp Physiol 77:761–764

Berger UV, Hediger MA (2000) Distribution of the glutamate transporters GLAST and GLT-1 in rat circumventricular organs, meninges, and dorsal root ganglia. J Comp Neurol 421:385–399

Bickerton RK, Buckley JO (1961) Evidence for a central mechanism in angiotensin-induced hypertension. Proc. Soc. Exp. Biol. Med. 106:834–836

Bishop VS, Sanderford MG (2000) Angiotensin II modulation of the arterial baroreflex: role of the area postrema. Clin Exp Pharmacol Physiol 27:428–431

Bisley JW, Rees SM, McKinley MJ, Hards DK, Oldfield BJ (1996) Identification of osmoresponsive neurons in the forebrain of the rat: a Fos study at the ultrastructural level. Brain Res 720:25–34

Blaine EH, Denton DA, McKinley MJ, Weller S (1975) A central osmosensitive receptor for renal sodium excretion. J Physiol 244:497–509

Blatteis CM, Bealer SL, Hunter WS, Llanos QJ, Ahokas RA, Mashburn TA, Jr. (1983) Suppression of fever after lesions of the anteroventral third ventricle in guinea pigs. Brain Res Bull 11:519–526

Blatteis CM, Hales JR, McKinley MJ, Fawcett AA (1987) Role of the anteroventral third ventricle region in fever in sheep. Can J Physiol Pharmacol 65:1255–1260

Blessing WW (1997) The Lower Brainstem and Bodily Homeostasis. Oxford University Press, New York

Blessing WW, Hedger SC, Joh TH, Willoughby JO (1987) Neurons in the area postrema are the only catecholamine-synthesizing cells in the medulla or pons with projections to the rostral ventrolateral medulla (C1-area) in the rabbit. Brain Res 419:336–340

Blurton-Jones MM, Roberts JA, Tuszynski MH (1999) Estrogen receptor immunoreactivity in the adult primate brain: neuronal distribution and association with p75, trkA, and choline acetyltransferase. J Comp Neurol 405:529–542

Boissonade FM, Sharkey KA, Davison JS (1994) Fos expression in ferret dorsal vagal complex after peripheral emetic stimuli. Am J Physiol 266:R1118–1126

Borison HL (1989) Area postrema: chemoreceptor circumventricular organ of the medulla oblongata. Prog Neurobiol 32:351–390

Borison HL, Brizzee KR (1951) Morphology of emetic chemoreceptor trigger zone in cat medulla oblongata. Proc Soc Exp Biol Med 77:38–42

Bouchaud C, Bosler O (1986) The circumventricular organs of the mammalian brain with special reference to monoaminergic innervation. Int Rev Cytol 105:283–327

Bouchaud C, Le Bert M, Dupouey P (1989) Are close contacts between astrocytes and endothelial cells a prerequisite condition of a blood-brain barrier? The rat subfornical organ as an example. Biol Cell 67:159–165
Bourque CW, Oliet SH, Kirkpatrick K, Richard D, Fisher TE (1993) Extrinsic and intrinsic modulatory mechanisms involved in regulating the electrical activity of supraoptic neurons. Ann N Y Acad Sci 689:512–519
Bourque CW, Oliet SH, Richard D (1994) Osmoreceptors, osmoreception, and osmoregulation. Front Neuroendocrinol 15:231–274
Brann DW, Mahesh VB (1995) Glutamate: a major neuroendocrine excitatory signal mediating steroid effects on gonadotropin secretion. J Steroid Biochem Mol Biol 53:325–329
Breder CD, Hazuka C, Ghayur T, Klug C, Huginin M, Yasuda K, Teng M, Saper CB (1994) Regional induction of tumor necrosis factor alpha expression in the mouse brain after systemic lipopolysaccharide administration. Proc Natl Acad Sci U S A 91:11393–11397
Brizzee KR, Klara PM (1984) The structure of the mammalian area postrema. Fed Proc 43:2944–2948
Brizzee KR, Neal LM (1954) A re-evaluation of the cellular morphology of the area postrema in view of recent evidence for a chemo-receptor function. J Comp Neurol 100:41–61
Brody MJ, Fink GD, Buggy J, Haywood JR, Gordon FJ, Johnson AK (1978) The role of the anteroventral third ventricle (AV3V) region in experimental hypertension. Circ Res 43, Suppl. I:I-2–13
Brownfield MS, Reid IA, Ganten D, Ganong WF (1982) Differential distribution of immunoreactive angiotensin and angiotensin- converting enzyme in rat brain. Neuroscience 7:1759–1769
Buggy J, Fisher AE (1977) Anteroventral third ventricle site of action for angiotensin induced thirst. Pharmacol Biochem Behav 4:651–660
Buggy J, Johnson AK (1978) Angiotensin-induced thirst: effects of third ventricle obstruction and periventricular ablation. Brain Res 149:117–128
Bunnemann B, Iwai N, Metzger R, Fuxe K, Inagami T, Ganten D (1992) The distribution of angiotensin II AT1 receptor subtype mRNA in the rat brain. Neurosci Lett 142:155–158
Buranarugsa P, Hubbard JI (1979) The neuronal organization of the rat subfornical organ in vitro and a test of the osmo- and morphine-receptor hypotheses. J Physiol 291:101–116
Butcher LL (1995) Cholinergic neurons and networks. In: Paxinos G (ed) The Rat Nervous System, Second Edition. Academic Press, San Diego, pp 1003–1016
Camacho A, Phillips MI (1981) Horseradish peroxidase study in rat of the neural connections of the organum vasculosum of the lamina terminalis. Neurosci Lett 25:201–204
Cao C, Matsumura K, Yamagata K, Watanabe Y (1996) Endothelial cells of the rat brain vasculature express cyclooxygenase-2 mRNA in response to systemic interleukin-1 beta: a possible site of prostaglandin synthesis responsible for fever. Brain Res 733:263–272
Carpenter DO (1990) Neural mechanisms of emesis. Can J Physiol Pharmacol 68:230–236
Carpenter DO, Briggs DB, Knox AP, Strominger N (1988) Excitation of area postrema neurons by transmitters, peptides, and cyclic nucleotides. J Neurophysiol 59:358–369
Carpenter DO, Briggs DB, Strominger N (1983) Responses of neurons of canine area postrema to neurotransmitters and peptides. Cell Mol Neurobiol 3:113–126
Ceccatelli S, Lundberg JM, Fahrenkrug J, Bredt DS, Snyder SH, Hokfelt T (1992) Evidence for involvement of nitric oxide in the regulation of hypothalamic portal blood flow. Neuroscience 51:769–772
Ceccatelli S, Villar MJ, Goldstein M, Hokfelt T (1989) Expression of c-Fos immunoreactivity in transmitter-characterized neurons after stress. Proc Natl Acad Sci U S A 86:9569–9573
Chai SY, Allen AM, Adam WR, Mendelsohn FA (1986) Local actions of angiotensin II: quantitative in vitro autoradiographic localization of angiotensin II receptor binding and angiotensin converting enzyme in target tissues. J Cardiovasc Pharmacol 8:S35-S39
Chai SY, McKenzie JS, McKinley MJ, Mendelsohn FA (1990) Angiotensin converting enzyme in the human basal forebrain and midbrain visualized by in vitro autoradiography. J Comp Neurol 291:179–194
Chai SY, McKinley MJ, Mendelsohn FA (1987) Distribution of angiotensin converting enzyme in sheep hypothalamus and medulla oblongata visualized by in vitro autoradiography. Clin Exp Hypertens A 9:449–460

Chai SY, McKinley MJ, Paxinos G, Mendelsohn FA (1991) Angiotensin converting enzyme in the monkey (Macaca fascicularis) brain visualized by in vitro autoradiography. Neuroscience 42:483–495
Charron G, Laforest S, Gagnon C, Drolet G, Mouginot D (2002) Acute sodium deficit triggers plasticity of the brain angiotensin type 1 receptors. FASEB J 16:610–612
Chernicky CL, Barnes KL, Conomy JP, Ferrario CM (1980) A morphological characterization of the canine area postrema. Neurosci Lett 20:37–43
Chernicky CL, Barnes KL, Ferrario CM, Conomy JP (1983) Brainstem distribution of neurons with efferent projections in the cervical vagus of the dog. Brain Res Bull 10:345–351
Chiaraviglio E, Perez Guaita MF (1984) Anterior third ventricle (A3 V) lesions and homeostasis regulation. J Physiol (Paris) 79:446–452
Ciriello J (1997) Afferent renal inputs onto subfornical organ neurons responsive to angiotensin II. Am J Physiol 272:R1684–1689
Ciriello J, Hochstenbach SL, Pastor Solano-Flores L (1996) Changes in NADPH diaphorase activity in forebrain structures of the laminae terminalis after chronic dehydration. Brain Res 708:167–172
Ciriello J, Hrycyshyn AW, Calaresu FR (1981) Glossopharyngeal and vagal afferent projections to the brain stem of the cat: a horseradish peroxidase study. J Auton Nerv Syst 4:63–79
Ciriello J, Rosas-Arellano MP, Solano-Flores LP (1996) Direct projections to subfornical organ from catecholaminergic neurons in the caudal nucleus of the solitary tract. Brain Res 726:227–232
Colombari DS, Menani JV, Johnson AK (1996) Forebrain angiotensin type 1 receptors and parabrachial serotonin in the control of NaCl and water intake. Am J Physiol 271:R1470–1476
Contreras RJ, Beckstead RM, Norgren R (1982) The central projections of the trigeminal, facial, glossopharyngeal and vagus nerves: an autoradiographic study in the rat. J Auton Nerv Syst 6:303–322
Cox BF, Bishop VS (1991) Neural and humoral mechanisms of angiotensin-dependent hypertension. Am J Physiol 261:H1284-H1291
Cunningham ET, Jr., Miselis RR, Sawchenko PE (1994) The relationship of efferent projections from the area postrema to vagal motor and brain stem catecholamine-containing cell groups: an axonal transport and immunohistochemical study in the rat. Neuroscience 58:635–648
Cunningham JT, Beltz T, Johnson RF, Johnson AK (1992) The effects of ibotenate lesions of the median preoptic nucleus on experimentally-induced and circadian drinking behavior in rats. Brain Res 580:325–330
Curtis KS, Huang W, Sved AF, Verbalis JG, Stricker EM (1999) Impaired osmoregulatory responses in rats with area postrema lesions. Am J Physiol 277:R209–219
Dampney RA (1994) Functional organization of central pathways regulating the cardiovascular system. Physiol Rev 74:323–364
Day HE, McKnight AT, Poat JA, Hughes J (1994) Evidence that cholecystokinin induces immediate early gene expression in the brainstem, hypothalamus and amygdala of the rat by a CCKA receptor mechanism. Neuropharmacology 33:719–727
Dellmann HD (1987) Fine structure. In: Gross PM (ed) Circumventricular Organs and Body Fluids. CRC Press, Boca Raton, pp 3–25 vol I
Dellmann HD (1998) Structure of the subfornical organ: a review. Microsc Res Tech 41:85–97
Dellmann HD, Simpson JB (1976) Regional differences in the morphology of the rat subfornical organ. Brain Res 116:389–400
Dellmann HD, Simpson JB (1979) The subfornical organ. Int Rev Cytol 58:333–421
Dempsey EW (1968) Fine structure of the rat's intercolumnar tubercle and its adjacent ependyma and chorioid plexus, with especial reference to the appearance of its sinusoidal vessels in experimental argyria. Exp Neurol 22:568–589
Dempsey EW (1973) Neural and vascular ultrastructure of the area postrema in the rat. J Comp Neurol 150:177–199
Denton DA, McKinley MJ, Weisinger RS (1996) Hypothalamic integration of body fluid regulation. Proc Natl Acad Sci U S A 93:7397–7404

Dourish CT, Duggan JA, Banks RJ (1992) Drinking induced by subcutaneous injection of angiotensin II in the rat is blocked by the selective AT1 receptor antagonist DuP 753 but not by the selective AT2 receptor antagonist WL 19. Eur J Pharmacol 211:113–116
Dournaud P, Slama A, Beaudet A, Epelbaum J (2000) Somatostatin receptors. In: Quirion R, Bjorklund A, Hokfelt T (eds) Handbook of Chemical Neuroanatomy. Peptide Receptors part I. Elsevier, Amsterdam.pp 1–44
Dragunow M, Faull R (1989) The use of c-fos as a metabolic marker in neuronal pathway tracing. J Neurosci Methods 29:261–265
Dumont Y, Jacques D, St Pierre J-A, Tong Y, Parker R, Herzog H, Quirion R (2000) Neuropeptide Y, peptide YY and pancreatic polypeptide receptor proteins and mRNAs in mammalian brains CNS. In: Quirion R, Bjorklund A, Hokfelt T (eds) Handbook of Chemical Neuroanatomy. Peptide Receptors part I. Elsevier, Amsterdam, pp 375–476
Duvernoy H, Koritke JG (1964) Contribution a l'etude de l'angioarchitectonic des organes circumventriculaires. Arch. Biol. (Liege) 75:849–904
Duvernoy H, Koritke JG (1965) Recherches sur la vascularisation de l'organe subfornical. J de med Besancon 2:115–130
Duvernoy H, Koritke JG (1969) Concerning the relationships of the circumventricular organs and their vessels with the cavity of the ventricles. In: Sterba G (ed) Zirkumventrikulare Organe und Liquor. VEB Gustav Fischer Verlag, Jena, pp 113–115
Duvernoy H, Koritke JG, Monnier G (1969) Sur la vascularisation de la lame terminale humaine. Z. Zellforsch. 102:49–77
Ehrlich KJ, Fitts DA (1990) Atrial natriuretic peptide in the subfornical organ reduces drinking induced by angiotensin or in response to water deprivation. Behav Neurosci 104:365–372
Ek M, Arias C, Sawchenko P, Ericsson-Dahlstrand A (2000) Distribution of the EP3 prostaglandin E(2) receptor subtype in the rat brain: relationship to sites of interleukin-1-induced cellular responsiveness. J Comp Neurol 428:5–20
Elmquist JK, Ackermann MR, Register KB, Rimler RB, Ross LR, Jacobson CD (1993) Induction of Fos-like immunoreactivity in the rat brain following Pasteurella multocida endotoxin administration. Endocrinology 133:3054–3057
Elmquist JK, Scammell TE, Jacobson CD, Saper CB (1996) Distribution of Fos-like immunoreactivity in the rat brain following intravenous lipopolysaccharide administration. J Comp Neurol 371:85–103
Emson PC, Goedert M, Mantyh PW (1985) Neurotensin-containing neurons. In: Bjorklund A, Hokfelt T (eds) Handbook of Chemical Neuroanatomy. Vol. 4. GABA and Neuropeptides in the CNS. Part I. Elsevier, Amsterdam, pp 355–405
Ericsson A, Kovacs KJ, Sawchenko PE (1994) A functional anatomical analysis of central pathways subserving the effects of interleukin-1 on stress-related neuroendocrine neurons. J Neurosci 14:897–913
Eriksson L, Fernandez O, Olsson K (1971) Differences in the antidiuretic response to intracarotid infusions of various hypertonic solutions in the conscious goat. Acta Physiol Scand 83:554–562
Felix D (1976) Peptide and acetylcholine action on neurons of the cat subfornical organ. Naunyn Schmiedebergs Arch Pharmacol 292:15–20
Felix H, Felix D, Sandri C, Akert K (1977) The surface morphology of the cat subfornical organ. In: Sterba G, Bargmann W (eds) Circumventriculare Organe. Deutsche Akademie der Naturforscher Leopoldina, Halle, pp 173–176
Ferguson AV, Bains JS (1996) Electrophysiology of the circumventricular organs. Front Neuroendocrinol 17:440–475
Ferguson AV, Day TA, Renaud LP (1984) Subfornical organ stimulation excites paraventricular neurons projecting to dorsal medulla. Am J Physiol 247:R1088–R1092
Ferguson AV, Marcus P (1988) Area postrema stimulation induced cardiovascular changes in the rat. Am J Physiol 255:R855-R860
Ferguson AV, Renaud LP (1984) Hypothalamic paraventricular nucleus lesions decrease pressor responses to subfornical organ stimulation. Brain Res 305:361–364
Ferguson AV, Renaud LP (1987) Neurophysiology of hypothalamic connections. In: Gross PM (ed) Circumventricular Organs and Body Fluids. CRC Press, Boca Raton, pp 43–58 vol I

Ferrario CM (1983) Central nervous system mechanisms of blood pressure control in normotensive and hypertensive states. Chest 83:331–335
Ferrario CM, Barnes KL, Diz DI, Block CH, Averill DB (1987) Role of area postrema pressor mechanisms in the regulation of arterial pressure. Can J Physiol Pharmacol 65:1591–1597
Ferrario CM, Barnes KL, Szilagyi JE, Brosnihan KB (1979) Physiological and pharmacological characterization of the area postrema pressor pathways in the normal dog. Hypertension 1:235–245
Ferrario CM, Dickinson CJ, McCubbin JW (1970) Central vasomotor stimulation by angiotensin. Clin Sci 39:239–245
Ferrario CM, Gildenberg PL, McCubbin JW (1972) Cardiovascular effects of angiotensin mediated by the central nervous system. Circ Res 30:257–262
Fink GD, Bruner CA, Pawloski CM, Blair ML, Skoog KM, Mangiapane ML (1986) Role of the area postrema in hypertension after unilateral renal artery constriction in the rat. Fed Proc 45:875
Fink GD, Bruner C, Mangiapane ML (1987) Area postrema is critical for angiotensin-induced hypertension in rats. Hypertension 9:355–361
Fitts DA, Masson DB (1990) Preoptic angiotensin and salt appetite. Behav Neurosci 104:643–650
Fitts DA, Starbuck EM, Ruhf A (2000) Circumventricular organs and ANG II-induced salt appetite: blood pressure and connectivity. Am J Physiol 279:R2277-R2286
Fitzsimons JT (1963) The effects of slow infusions of hypertonic solutions on drinking and drinking thresholds in rats. J Physiol (Lond) 167:344–354
Fitzsimons JT, Kucharczyk J (1978) Drinking and haemodynamic changes induced in the dog by intracranial injection of components of the renin-angiotensin system. J Physiol (Lond) 276:419–434
Fuxe K, Owman C (1965) Cellular localisation of monoamines in the area postrema of certain mammals. J. Comp. Neurol. 125:337–354
Gardiner TW, Stricker EM (1985) Impaired drinking responses of rats with lesions of nucleus medianus: circadian dependence. Am J Physiol 248:R224-R230
Gardiner TW, Verbalis JG, Stricker EM (1985) Impaired secretion of vasopressin and oxytocin in rats after lesions of nucleus medianus. Am J Physiol 249:R681–688
Gebke E, Muller AR, Pehl U, Gerstberger R (2000) Astrocytes in sensory circumventricular organs of the rat brain express functional binding sites for endothelin. Neuroscience 97:371–381
Geddes BJ, Parry LJ, Summerlee AJ (1994) Brain angiotensin-II partially mediates the effects of relaxin on vasopressin and oxytocin release in anesthetized rats. Endocrinology 134:1188–1192
Geddes BJ, Summerlee AJ (1995) The emerging concept of relaxin as a centrally acting peptide hormone with hemodynamic actions. J Neuroendocrinol 7:411–417
Giles ME, Fernley RT, Nakamura Y, Moeller I, Aldred GP, Ferraro T, Penschow JD, McKinley MJ, Oldfield BJ (1999) Characterization of a specific antibody to the rat angiotensin II AT_1 receptor. J Histochem Cytochem 47:507–516
Giles ME, Sly DJ, McKinley MJ, Oldfield BJ (2001) Neurons in the lamina terminalis which project polysynaptically to the kidney express angiotensin AT_{1A} receptor. Brain Res 898:9–12
Gilman A (1937) The relation between blood osmotic pressure, fluid distribution and voluntary water intake. Am J Physiol 120:323–328
Göke R, Larsen PJ, Mikkelsen JD, Sheikh SP (1995) Distribution of GLP-1 binding sites in the rat brain: evidence that exendin-4 is a ligand of brain GLP-1 binding sites. Eur J Neurosci 7:2294–2300
Grafe G, Weindl A (1987) The vascular connections of the organum vasculosum of the lamina terminalis in the rat. Wiss Z Karl Marx-Univ Leipzig, Math-Naturwiss R 36:214–220
Gross PM (1991) Morphology and physiology of capillary systems in subregions of the subfornical organ and area postrema. Can J Physiol Pharmacol 69:1010–25
Gross PM, ed. (1987) Circumventricular Organs and Body Fluids. Vol I-III. CRC Press, Boca Raton
Gu GB, Ju G (1995) The parabrachio-subfornical organ projection in the rat. Brain Res Bull 38:41–47
Gu GB, Simerly RB (1997) Projections of the sexually dimorphic anteroventral periventricular nucleus in the female rat. J Comp Neurol 384:142–164

Guan JL, Wang QP, Shioda S (2000) Observation of the ultrastructure and synaptic relationships of angiotensin II-like immunoreactive neurons in the rat area postrema. Synapse 38:231–237
Guldenaar SE, Noctor SC, McCabe JT (1992) Fos-like immunoreactivity in the brain of homozygous diabetes insipidus Brattleboro and normal Long-Evans rats. J Comp Neurol 322:439–448
Gutman MB, Ciriello J, Mogenson GJ (1988) Effects of plasma angiotensin II and hypernatremia on subfornical organ neurons. Am J Physiol 254:R746-R754
Hamamura M, Nunez DJ, Leng G, Emson PC, Kiyama H (1992) c-fos may code for a common transcription factor within the hypothalamic neural circuits involved in osmoregulation. Brain Res 572:42–51
Han L, Rowland NE (1995) Sodium depletion and Fos-immunoreactivity in lamina terminalis. Neurosci Lett 193:173–176
Hashimoto M, Ueno T, Iriki M (1994) What roles does the organum vasculosum laminae terminalis play in fever in rabbits? Pflugers Arch 429:50–57
Hasser EM, Cunningham JT, Sullivan MJ, Curtis KS, Blaine EH, Hay M (2000) Area postrema and sympathetic nervous system effects of vasopressin and angiotensin II. Clin Exp Pharmacol Physiol 27:432–436
Hasser EM, Nelson DO, Haywood JR, Bishop VS (1987) Inhibition of renal sympathetic nervous activity by area postrema stimulation in rabbits. Am J Physiol 253:H91-H99
Hattori Y, Kasai M, Uesugi S, Kawata M, Yamashita H (1988) Atrial natriuretic polypeptide depresses angiotensin II induced excitation of neurons in the rat subfornical organ *in vitro*. Brain Res 443:355–359
Hay M, Lindsley KA (1999) AMPA receptor activation of area postrema neurons. Am J Physiol 276:R586–90
Hay M, McKenzie H, Lindsley K, Dietz N, Bradley SR, Conn PJ, Hasser EM (1999) Heterogeneity of metabotropic glutamate receptors in autonomic cell groups of the medulla oblongata of the rat. J Comp Neurol 403:486–501
Haywood JR, Fink GD, Buggy J, Phillips MI, Brody MJ (1980) The area postrema plays no role in the pressor action of angiotensin in the rat. Am J Physiol 239:H108-H113
Henke H, Tobler PH, Fischer JA (1983) Localization of salmon calcitonin binding sites in rat brain by autoradiography. Brain Res 272:373–377
Herbert H, Saper CB (1990) Cholecystokinin-, galanin-, and corticotropin-releasing factor-like immunoreactive projections from the nucleus of the solitary tract to the parabrachial nucleus in the rat. J Comp Neurol 293:581–598
Herbert H, Moga MM, Saper CB (1991) Connections of the parabrachial nucleus with the nucleus of the solitary tract and the medullary reticular formation of the rat. J. Comp. Neurol. 293:540–580
Herbert J, Forsling ML, Howes SR, Stacey PM, Shiers HM (1992) Regional expression of c-fos antigen in the basal forebrain following intraventricular infusions of angiotensin and its modulation by drinking either water or saline. Neuroscience 51:867–882
Hernesniemi J, Kawana E, Bruppacher H, Sandri C (1972) Afferent connections of the subfornical organ and of the supraoptic crest. Acta Anat 81:321–336
Hill DR, Shaw TM, Graham W, Woodruff GN (1990) Autoradiographical detection of cholecystokinin-A receptors in primate brain using 125I-Bolton Hunter CCK-8 and 3H-MK-329. J Neurosci 10:1070–1081
Hill DR, Woodruff GN (1990) Differentiation of central cholecystokinin receptor binding sites using the non-peptide antagonists MK-329 and L-365,260. Brain Res 526:276–283
Hirunagi K, Ishikawa A, Namikawa T, Uryu K (1995) Immunocytochemical identification of serotonergic supraependymal nerve fibers in the third ventricle of the house musk shrew, Suncus murinus. Anat Anz 177:297–304
Hochstenbach SL, Ciriello J (1996) Effect of lesions of forebrain circumventricular organs on c-fos expression in the central nervous system to plasma hypernatremia. Brain Res 713:17–28
Hofer H (1958) Zur Morphologie der circumventrikularen Organe des Zwischenhirnes der Saugetiere. Vorh. dtsch. zool. Ges. Frankfort/M 8:22:202–251
Hofer H (1969) Zur Anatomie der circumventricular Organe. In: Sterba G (ed) Zirkumventrikulare Organe und Liquor. VEB Gustav Fischer Verlag, Jena, pp 77–88

Hoffman GE, Lee WS, Attardi B, Yann V, Fitzsimmons MD (1990) Luteinizing hormone-releasing hormone neurons express c-fos antigen after steroid activation. Endocrinology 126:1736–1741

Holmes JH, Gregersen MI (1950) Observations on drinking induced by hypertonic solutions. Am J Physiol 162:326–337

Hori T, Kaizuka Y, Takaki A, Katafuchi T (2001) Thermal stress and immunity. In: Kosaka M, Sugahara T, Schmidt KL, Simon E (eds) Thermotherapy for Neoplasia, Inflammation, and Pain. Springer-Verag, Tokyo pp 242–251

Hori T, Kiyohara T, Nakashima T, Mizuno K, Muratani H, Katafuchi T (1992) Effects of temperature and neuroactive substances on hypothalamic neurones in vitro: possible implications for the induction of fever. Physiol Res 41:77–81

Huang W, Lee SL, Sjoquist M (1995) Natriuretic role of endogenous oxytocin in male rats infused with hypertonic NaCl. Am J Physiol 268:R634-R640

Huang W, Sved A, Stricker EM (2000) Vasopressin and oxytocin release evoked by NaCl loads are selectively blunted by area postrema lesions. Am J Physiol 278:R732-R740

Hubschle T, McKinley MJ, Oldfield BJ (1998) Efferent connections of the lamina terminalis, the preoptic area and the insular cortex to submandibular and sublingual gland of the rat traced with pseudorabies virus. Brain Res 806:219–231

Hunt SP, Pini A, Evan G (1987) Induction of c-fos-like protein in spinal cord neurons following sensory stimulation. Nature 328:632–634

Hyde TM, Miselis RR (1984) Area postrema and adjacent nucleus of the solitary tract in water and sodium balance. Am J Physiol. 247:R173-R182

Jahn K, Bufler J, Weindl A, Arzberger T, Hatt H (1996) Patch-clamp study on membrane properties and transmitter activated currents of rabbit area postrema neurons. J Comp Physiol [A] 178:771–778

Jewell PA, Verney EB (1957) An experimental attempt to determine the site of the neurohypohysial osmoreceptors in the dog. Phil Trans Roy Soc B 240B

Johnson AK, Buggy J (1978) Periventricular preoptic-hypothalamus is vital for thirst and normal water economy. Am J Physiol 234:R122–R129

Johnson AK, Gross PM (1993) Sensory circumventricular organs and brain homeostatic pathways. FASEB J 7:678–686

Johnson AK, Hoffman WE, Buggy J (1978) Attenuated pressor responses to intracranially injected stimuli and altered antidiuretic activity following preoptic-hypothalamic periventricular ablation. Brain Res 157:161–166

Johnson AK, Mann JF, Rascher W, Johnson JK, Ganten D (1981) Plasma angiotensin II concentrations and experimentally induced thirst. Am J Physiol 240:R229-R234

Johnson AK, Wilkin LD (1987) The lamina terminalis. In: Gross PM (ed) Circumventricular Organs and Body Fluids. CRC Press, Boca Raton, pp 125–141 vol III)

Johnson RF, Beltz TG, Sharma RV, Xu Z, Bhatty RA, Johnson AK (2001) Agonist activation of cytosolic Ca^{2+} in subfornical organ cells projecting to the supraoptic nucleus. Am J Physiol Regul Integr Comp Physiol 280:R1592–1599

Joy MD, Lowe RD (1970) Evidence that the area postrema mediates the central cardiovascular response to angiotensin II. Nature 228:1303–1304

Jurzak M, Schmid H, Gerstberger R (1994) NADPH-diaphorase staining and NO-synthase immunoreactivity in circumventricular organs of the rat brain. In: Pleschka K, Gerstberger R (eds) Integrative and Cellular aspects of autonomic functions: temperature and osmoregulation. John Libbey Eurotext, Paris, pp 451–459

Kalia M, Mesulam MM (1980) Brain stem projections of sensory and motor components of the vagus complex in the cat: I. The cervical vagus and nodose ganglion. J Comp Neurol 193:435–465

Kalia M, Sullivan JM (1982) Brainstem projections of sensory and motor components of the vagus nerve in the rat. J Comp Neurol 211:248–265

Katsuura G, Arimura A, Koves K, Gottschall PE (1990) Involvement of organum vasculosum of lamina terminalis and preoptic area in interleukin 1 beta-induced ACTH release. Am J Physiol 258:E163–171

Kawano H, Masuko S (2001) Tyrosine hydroxylase-immunoreactivite projections from the caudal ventrolateral medulla to the subfornical organ in the rat. Brain Res 903:154–161

Khachaturian H, Lewis ME, Tsao K, Watson SJ (1985) Beta-endorphin, alpha MSH, ACTH, and related peptides. In: Bjorklund A, Hokfekt T (eds) Handbook of Chemical Neuroanatomy. Vol. 4. GABA and Neuropeptides in the CNS. Part I. Elsevier, Amsterdam, pp 216–272

King JC, Tobet SA, Snavely FL, Arimura AA (1982) LHRH immunopositive cells and their projections to the median eminence and organum vasculosum of the lamina terminalis. J Comp Neurol 209:287–300

Klara PM, Brizzee KR (1975) The ultrastructural morphology of the squirrel monkey area postrema. Cell Tissue Res 160:315–326

Klara PM, Brizzee KR (1977) Ultrastructure of the feline area postrema. J Comp Neurol 72:409–431

Kohzuki M, Chai SY, Paxinos G, Karavas A, Casley DJ, Johnston CI, Mendelsohn FA (1991) Localization and characterization of endothelin receptor binding sites in the rat brain visualized by in vitro autoradiography. Neuroscience 42:245–260

Kovacs KJ, Sawchenko PE (1993) Mediation of osmoregulatory influences on neuroendocrine corticotropin- releasing factor expression by the ventral lamina terminalis. Proc Natl Acad Sci U S A 90:7681–685

Krisch B (1986) The functional and structural borders between the CSF- and blood- dominated milieus in the choroid plexuses and the area postrema of the rat. Cell Tissue Res 245:101–115

Krisch B (1992) Somatostatin-binding sites on structures of circumventricular organs. Prog Brain Res 91:247–250

Krisch B, Leonhardt H, Buchheim W (1978) The functional and structural border between the CSF- and blood-milieu in the circumventricular organs (organum vasculosum laminae terminalis, subfornical organ, area postrema) of the rat. Cell Tissue Res 195:485–497

Krisch B, Leonhardt H, Oksche A (1987) Compartments in the organum vasculosum laminae terminalis of the rat and their delineation against the outer cerebrospinal fluid-containing space. Cell Tissue Res 250:331–347

Kroidl R (1968) Die arterielle una venose versorgung der area postrema der ratte. Z Zellforsch 89:430–452

Krout KE, Kawano J, Mettenleiter TC, Loewy AD (2001) CNS inputs to the suprachiasmatic nucleus of the rat. Neurosci 110:73–92

Krukoff TL, Vu T, Harris KH, Aippersbach S, Jhamandas JH (1992) Neurons in the rat medulla oblongata containing neuropeptide Y-, angiotensin II-, or galanin-like immunoreactivity project to the parabrachial nucleus. Neuroscience 47:175–184

Kuhlenbeck H (1970) The central nervous system of vertebrates. Vol. 3. Part 1. Structure elements; biology of nervous tissue. Karger, Basel

Kunii K, Yamanaka A, Nambu T, Matsuzaki I, Goto K, Sakurai T (1999) Orexins/hypocretins regulate drinking behaviour. Brain Res 842:256–261

Kurokawa K, Yamada H, Ochi J (1997) Topographical distribution of neurons containing endothelin type A receptor in the rat brain. J Comp Neurol 389:348–360

Lacassagne O, Kessler JP (2000) Cellular and subcellular distribution of the amino-3-hydroxy-5-methyl-4- isoxazole propionate receptor subunit GluR2 in the rat dorsal vagal complex. Neuroscience 99:557–563

Lacroix S, Feinstein D, Rivest S (1998) The bacterial endotoxin lipopolysaccharide has the ability to target the brain in upregulating its membrane CD14 receptor within specific cellular populations. Brain Pathol 8:625–640

Lacroix S, Rivest S (1997) Functional circuitry in the brain of immune challenged rats: partial involvement of prostaglandins. J Comp Neurol 387:307–324

Laflamme N, Rivest S (2001) Toll-like receptor 4: the missing link of the cerebral innate immune response triggered by circulating gram-negative bacterial cell wall components. FASEB J 15:155–163

Lanca AJ, van der Kooy D (1985) A serotonin-containing pathway from the area postrema to the parabrachial nucleus in the rat. Neuroscience 14:1117–1126

Landas S, Phillips I (1987) Comparative anatomy of the organum vasculosum of the lamina terminalis. In: Gross PM (ed) Circumventricular Organs and Body Fluids. CRC Press, Boca Raton, pp 131–156 vol I

Larsen PJ, Mikkelsen JD (1995) Functional identification of central afferent projections conveying information of acute “stress” to the hypothalamic paraventricular nucleus. J Neurosci 15:2609–2627

Larsen PJ, Moller M, Mikkelsen JD (1991) Efferent projections from the periventricular and medial parvicellular subnuclei of the hypothalamic paraventricular nucleus to circumventricular organs of the rat: a Phaseolus vulgaris- leucoagglutinin (PHA-L) tracing study. J Comp Neurol 306:462–479

Le Beux YJ (1972) An ultrastructural study of the neurosecretory cells of the medial vascular prechiasmatic gland. II. Nerve endings. Z. Zellforsch. 127:439–461

Lebel E, Vallieres L, Rivest S (2000) Selective involvement of interleukin-6 in the transcriptional activation of the suppressor of cytokine signaling-3 in the brain during systemic immune challenges. Endocrinology 141:3749–3763

Lee HS, Chong W, Han SK, Lee MH, Ryu PD (2001) Activation of metabotropic glutamate receptors inhibits gabaergic transmission in the rat subfornical organ. Neurosci 102:401–411

Lee HY, Whiteside MB, Herkenham M (1998) Area postrema removal abolishes stimulatory effects of intravenous interleukin-1 beta on hypothalamic-pituitary-adrenal axis activity and c-fos mRNA in the hypothalamic paraventricular nucleus. Brain Res Bull 46:495–503

Leng G, Blackburn RE, Dyball REJ, Russell JA (1989) Role of the anterior periventricular structures in the regulation of supraoptic neuronal activity and neurohypohysial hormone secretion in the rat. J. Neuroendocrinol 1:35–46

Lenkei Z, Corvol P, Llorens-Cortes C (1995) The angiotensin receptor subtype AT1A predominates in rat forebrain areas involved in blood pressure, body fluid homeostasis and neuroendocrine control. Brain Res Mol Brain Res 30:53–60

Lenkei Z, Palkovits M, Corvol P, Llorens-Cortes C (1997) Expression of angiotensin type-1 (AT1) and type-2 (AT2) receptor mRNAs in the adult rat brain: a functional neuroanatomical review. Front Neuroendocrinol 18:383–439

Leonhardt H (1980) Ependym und circumventriculare Organe. In: Oksche A, Vollrath L (eds) Handbuch der mikroscopischen Anatomie des Menschen. Springer Verlag, Berlin, pp 177–666

Leonhardt H, Lindemann B (1973) Surface morphology of the subfornical organ in the rabbit’s brain. Z Zellforsch Mikrosk Anat 146:243–260

Lescure H, Dufy B, Leonardelli J, Bensch C (1978) Organum vasculosum laminae terminalis and reflex ovulation in the rabbit. Brain Res 154:209–213

Leslie RA (1986) Comparative aspects of the area postrema: fine-structural considerations help to determine its function. Cell Mol Neurobiol 6:95–120

Leveque TF, Stutinsky F, Porte A, Stoeckel M-E (1967) Ultrastructure of the prechiasmatic gland in the rat and mouse. Neuroendocrinol 2:56–63

Lewis PR, Shute CCD (1967) The cholinergic limbic system; projections to hippocampal formation, medial cortex, nuclei of the ascending cholinergic system, and the subfornical organ and supra-optic crest. Brain 90:521–540

Li YW, Dampney RA (1994) Expression of Fos-like protein in brain following sustained hypertension and hypotension in conscious rabbits. Neuroscience 61:613–634

Li BH, Rowland NE (1995) Effects of vagotomy on cholecystokinin- and dexfenfluramine induced Fos-like immunoreactivity in the rat brain. Brain Res Bull 37:589–593

Liedtke W, Choe Y, Marti-Renom MA, Bell AM, Denis CS, Sali A, Hudspeth AJ, Friedman JM, Heller S (2000) Vanilloid receptor-related osmotically activated channel (VR-OAC), a candidate vertebrate osmoreceptor. Cell 103:525–535

Lin MT, Lin JH (2000) Involvement of tyrosine kinase in the pyrogenic fever exerted by NOS pathways in organum vasculosum laminae terminalis. Neuropharmacol 39:347–352

Lin MT, Pan SP, Lin JH, Yang YL (1999) Central control of blood pressure by nitrergic mechanisms in organum vasculosum laminae terminalis of rat brain. Br J Pharmacol 127:1511–1517

Lind RW (1986) Bi-directional, chemically specified neural connections between the subfornical organ and the midbrain raphe system. Brain Res 384:250–261

Lind RW (1987) Neural connections. In: Gross PM (ed) Circumventricular Organs and Body Fluids. CRC Press, Boca Raton, pp 27–42 vol I
Lind RW, Swanson LW, Ganten D (1984) Angiotensin II immunoreactivity in the neural afferents and efferents of the subfornical organ of the rat. Brain Res 321:209–215
Lind RW, Swanson LW, Ganten, D (1985) Organization of angiotensin II immunoreactive cells and fibers in the rat central nervous system. An immunohistochemical study. Neuroendocrinol 40:2–24
Lind RW, Swanson LW, Ganten D (1984) Angiotensin II immunoreactive pathways in the central nervous system of the rat: evidence for a projection from the subfornical organ to the paraventricular nucleus of the hypothalamus. Clin Exp Hypertens A 6:1915–1920
Lind RW, Van Hoesen GW, Johnson AK (1982) An HRP study of the connections of the subfornical organ of the rat. J Comp Neurol 210:265–277
Lindheimer MD, Barron WM, Davison JM (1989) Osmoregulation of thirst and vasopressin release in pregnancy. Am J Physiol 257:F159F-F1569
Lindstrom PA, Brizzee KR (1962) Relief of intractable vomiting from surgical lesions in the area postrema. J Neurosurg 19:228–236
Lowe RD, Scroop GC (1969) The cardiovascular response to vertebral artery infusions of angiotensin in the dog. Clin Sci 37:593–603
Luckman SM (1997) Comparison of the expression of c-fos, nur77 and egr1 mRNAs in rat hypothalamic magnocellular neurons and their putative afferent projection neurons: cell- and stimulus-specific induction. Eur J Neurosci 9:2443–2451
Luckman SM, Hamamura M, Antonijevic I, Dye S, Leng G (1993) Involvement of cholecystokinin receptor types in pathways controlling oxytocin secretion. Br J Pharmacol 110:378–84
Luckman SM, Huckett L, Bicknell RJ, Voisin DL, Herbison AE (1997) Up-regulation of nitric oxide synthase messenger RNA in an integrated forebrain circuit involved in oxytocin secretion. Neuroscience 77:37–48
Lynch KR, Hawelu-Johnson CL, Guyenet PG (1987) Localization of brain angiotensinogen mRNA by hybridization histochemistry. Brain Res 388:149–158
Maley BE, Newton BW, Howes KA, Herman LM, Oloff CM, Smith KC, Elde RP (1987) Immunohistochemical localization of substance P and enkephalin in the nucleus tractus solitarii of the rhesus monkey, Macaca mulatta. J Comp Neurol 260:483–490
Mangiapane ML (1987) Cardiovascular regulation. In: Gross PM (ed) Circumventricular Organs and Body Fluids. CRC Press, Boca Raton, pp 95–108 vol I
Mangiapane ML, Brody MJ (1987) Vasoconstrictor and vasodilator sites within anteroventral third ventricle region. Am J Physiol 253:R827-R831
Mangiapane ML, Simpson JB (1980) Subfornical organ lesions reduce the pressor effect of systemic angiotensin II. Neuroendocrinology 31:380–384
Mangiapane ML, Simpson JB (1983) Drinking and pressor responses after acetylcholine injection into subfornical organ. Am J Physiol 244:R508-R513
Mangiapane ML, Thrasher TN, Keil LC, Simpson JB, Ganong WF (1983) Deficits in drinking and vasopressin secretion after lesions of the nucleus medianus. Neuroendocrinology 37:73–77
Mangiapane ML, Thrasher TN, Keil LC, Simpson JB, Ganong WF (1984) Role for the subfornical organ in vasopressin release. Brain Res Bull 13:43–47
Mann JF, Johnson AK, Ganten D (1980) Plasma angiotensin II: dipsogenic levels and angiotensin-generating capacity of renin. Am J Physiol 238:R372-R377
Mark MH, Farmer PM (1984) The human subfornical organ: an anatomic and ultrastructural study. Ann Clin Lab Sci 14:427–442
Mason WT (1980) Supraoptic neurones of rat hypothalamus are osmosensitive. Nature 287:154–157
Matsuda T, Hori T, Nakashima T (1992) Thermal and PGE2 sensitivity of the organum vasculosum lamina terminalis region and preoptic area in rat brain slices. J Physiol 454:197–212
Matsumura K, Kobayashi S (2001) Neuroanatomy of fever: localization of cytokine and prostaglandin systems in the brain. In: Kosaka M, Sugahara T, Schmidt KL, Simon E (eds) Thermotherapy for Neoplasia, Inflamation and Pain. Springer-Verlag, Tokyo, pp 290–299

Matsumura K, Watanabe Y, Onoe H, Hayaishi O (1990) High density of prostaglandin E2 binding sites in the anterior wall of the 3rd ventricle: a possible site of its hyperthermic action. Brain Res 533:147–151

May CN, McAllen RM, McKinley MJ (2000) Renal nerve inhibition by central NaCl and ANG II is abolished by lesions of the lamina terminalis. Am J Physiol Regul Integr Comp Physiol 279:R1827-R1833

McAllen RM, Pennington GL, McKinley MJ (1990) Osmoresponsive units in sheep median preoptic nucleus. Am J Physiol 259:R593-R600

McDonald TJ, Li C, Nijland MJ, Caston-Balderrama A, Ross MG (1998) Fos response of fetal sheep anterior circumventricular organs to osmotic challenge in late gestation. Am J Physiol 275:H609-H614

McKinley MJ (1985) Volume regulation of antidiuretic hormone secretion. In: Ganten D, Pfaff D (eds) Current Topics in Neuroendocrinology. Neurobiology of Vasopressin. Springer-Verlag, Berlin, pp 61–100

McKinley MJ, Allen A, Clevers J, Denton DA, Mendelsohn FA (1986a) Autoradiographic localization of angiotensin receptors in the sheep brain. Brain Res 375:373–376

McKinley MJ, Allen AM, Burns P, Colvill LM, Oldfield BJ (1998) Interaction of circulating hormones with the brain: the roles of the subfornical organ and the organum vasculosum of the lamina terminalis. Clin Exp Pharmacol Physiol Suppl 25:S61–S67

McKinley MJ, Allen AM, Clevers J, Paxinos G, Mendelsohn FA (1987a) Angiotensin receptor binding in human hypothalamus: autoradiographic localization. Brain Res 420:375–379

McKinley MJ, Badoer E, Oldfield BJ (1992a) Intravenous angiotensin II induces Fos-immunoreactivity in circumventricular organs of the lamina terminalis. Brain Res 594:295–300

McKinley MJ, Badoer E, Vivas L, Oldfield BJ (1995) Comparison of c-fos expression in the lamina terminalis of conscious rats after intravenous or intracerebroventricular angiotensin. Brain Res Bull 37:131–137

McKinley MJ, Badoer E, Weisinger RS, Oldfield BJ (1992b) C-fos expression induced in the circumventricular organs of the lamina terminalis by circulating angiotensin II in rats. Abstracts Soc. Neurosci. 18:143

McKinley MJ, Bicknell RJ, Hards D, McAllen RM, Vivas L, Weisinger RS, Oldfield BJ (1992c) Efferent neural pathways of the lamina terminalis subserving osmoregulation. Prog Brain Res 91:395–402

McKinley MJ, Burns P, Colvill LM, Oldfield BJ, Wade JD, Weisinger RS, Tregear GW (1997a) Distribution of Fos immunoreactivity in the lamina terminalis and hypothalamus induced by centrally administered relaxin in conscious rats. J Neuroendocrinol 9:431–437

McKinley MJ, Clevers J, Denton DA, Oldfield BJ, Penschow J, Rundgren M (1987b) Fine structure of the organum vasculosum of the lamina terminalis. In: Gross PM (ed) Circumventricular Organs and Body Fluids. CRC Press, Boca Raton, pp 111–130 vol I

McKinley MJ, Coghlan JP, Congiu M, Denton DA, Fei DT, Park RG (1986b) Augmented plasma renin levels in dehydrated sheep with periventricular lesions. Brain Res 376:416–419

McKinley MJ, Colvill LM, Giles ME, Oldfield BJ (1997b) Distribution of Fos-immunoreactivity in rat brain following a dipsogenic dose of captopril and effects of angiotensin receptor blockade. Brain Res 747:43–51

McKinley MJ, Congiu M, Denton DA, Park RG, Penschow J, Simpson JB, Tarjan E, Weisinger RS, Wright RD (1984) The anterior wall of the third cerebral ventricle and homeostatic response to dehydration. Journal de Physiologie (Paris) 79:421–427

McKinley MJ, Congiu M, Miselis RR, Oldfield BJ, Pennington G (1988) The lamina terminalis and osmotically stimulated vasopressin secretion. In: Yoshida S, Share L (eds) Recent Progress in Pituitary Hormones 1988. Excerpta Medica, Amsterdam, pp 117–124

McKinley MJ, Denton DA, Leksell LG, Mouw DR, Scoggins BA, Smith MH, Weisinger RS, Wright RD (1982) Osmoregulatory thirst in sheep is disrupted by ablation of the anterior wall of the optic recess. Brain Res 236:210–215

McKinley MJ, Denton DA, Leventer M, Miselis RR, Park RG, Tarjan E, Simpson JB, Weisinger RS (1986c) Adipsia in sheep caused by cerebral lesions. In: de Caro G, Massi M, Epstein AN (eds) The Physiology of Thirst and Sodium Appetite. Plenum Press, New York, pp 321–326

McKinley MJ, Denton DA, Leventer M, Penschow J, Weisinger RS, Wright RD (1983a) Morphology of the organum vasculosum of the lamina terminalis (OVLT) of the sheep. Brain Res Bull 11:649–657
McKinley MJ, Denton DA, Nelson JF, Weisinger RS (1983b) Dehydration induces sodium depletion in rats, rabbits and sheep. Am J Physiol 245:R287-R292
McKinley MJ, Denton DA, Park RG, Weisinger RS (1983c) Cerebral involvement in dehydration induced natriuresis. Brain Res 263:340–343
McKinley MJ, Denton DA, Park RG, Weisinger RS (1986d) Ablation of subfornical organ does not prevent angiotensin-induced water drinking in sheep. Am J Physiol 250:R1052–1059
McKinley MJ, Denton DA, Weisinger RS (1978) Sensors for antidiuresis and thirst-osmoreceptors or CSF sodium detectors? Brain Res 141:89–103
McKinley MJ, Evered MD, Mathai ML (2000) Renal Na excretion in dehydrated and rehydrated adrenalectomised sheep maintained with aldosterone. Am J Physiol 279: R17-R24
McKinley MJ, Giles M, Oldfield BJ (1996a) Regrowth and redirection of vasopressin- and oxytocin-containing fibers after ablation of the neurohypophyseal tract in sheep. Abstracts Soc Neurosci 22:1488
McKinley MJ, Hards DK, Oldfield BJ (1994a) Identification of neural pathways activated in dehydrated rats by means of Fos-immunohistochemistry and neural tracing. Brain Res 653:305–314
McKinley MJ, Lichardus B, McDougall JG, Weisinger RS (1992d) Periventricular lesions block natriuresis to hypertonic but not isotonic NaCl loads. Am J Physiol 262:F98–107
McKinley MJ, Mathai ML, Pennington G, Rundgren M, Vivas L (1999) Effect of individual or combined ablation of the nuclear groups of the lamina terminalis on water drinking in sheep. Am J Physiol 276:R673–683
McKinley MJ, McAllen RM, Mendelsohn FAO, Allen AM, Chai S-Y, Oldfield BJ (1990) Circumventricular organs: Neuroendocrine interfaces between the brain and the hemal milieu. Front-Neuroendocrinol 11:91–127
McKinley MJ, McAllen RM, Pennington GL, Smardencas A, Weisinger RS, Oldfield BJ (1996b) Physiological actions of angiotensin II mediated by AT1 and AT2 receptors in the brain. Clin Exp Pharmacol Physiol 3:S99-S104
McKinley MJ, McBurnie MI, Mathai ML (2001) Neural mechanisms subserving central angiotensinergic influences on plasma renin in sheep. Hypertension 37:1375–1381
McKinley MJ, Oldfield BJ (1994c) Distribution of Fos-like immunoreactivity in rat brain following isoproterenol treatment and the effects of converting enzyme inhibition or angiotensin II blockade. Abstracts Soc Neurosci 20:1679
McKinley MJ, Pennington G, Oldfield BJ (1996c) The anteroventral third ventricle and dorsal lamina terminalis: headquarters for body fluid homeostasis. Clin. Exp. Pharmcol. Physiol. 23:271–281
McKinley MJ, Rundgren M, Coghlan JP (1994d) Cerebral osmoregulatory reduction of plasma renin concentration in sheep. Acta Physiol Scand 152:323–332
Melander T, Kohler C, Nilsson S, Hokfelt T, Brodin E, Theodorsson E, Bartfai T (1988) Autoradiographic quantitation and anatomical mapping of 125I-galanin binding sites in the rat central nervous system. J Chem Neuroanat 1:213–233
Menani JV, Columbari DSA, Beltz TG, Thunhorst RL, Johnson AK (1998) Salt appetite: interaction of forebrain angiotensinergic and hindbrain serotonergic mechanisms. Brain Res 801:29–35
Mendelsohn FA, Allen AM, Clevers J, Denton DA, Tarjan E, McKinley MJ (1988) Localization of angiotensin II receptor binding in rabbit brain by in vitro autoradiography. J Comp Neurol 270:372–384
Mendelsohn FA, Quirion R, Saavedra JM, Aguilera G, Catt KJ (1984) Autoradiographic localization of angiotensin II receptors in rat brain. Proc Natl Acad Sci U S A 81:1575–1579
Mercer LD, Le VQ, Nunan J, Jones NM, Beart PM (2000) Direct visualization of cholecystokinin subtype 2 receptors in rat central nervous system using anti-peptide antibodies. Neurosci Lett 293:167–170
Mergner H (1959) Die Blutversorgung der Lamina terminalis bei einigen Agfen. Z wiss Zoology 165:140–185

Meszaros T, Leranth C, Palkovits M, Hazas J (1969) Secretory and esterase activity of the circumventricular organs with special reference to the infundibular ependyma. In: Sterba G (ed) Zirkumventrikulare Organe und Liquor. VEB Gustav Fischer Verlag, Jena, pp 131–134

Meyer U, Kruhoffer M, Flugge G, Fuchs E (1998) Cloning of glucocorticoid receptor and mineralocorticoid receptor cDNA and gene expression in the central nervous system of the tree shrew (Tupaia belangeri). Brain Res Mol Brain Res 55:243–253

Miller AD, Leslie RA (1994) The area postrema and vomiting. Front Neuroendocrinol 15:301–320

Miller AD, Ruggiero DA (1994) Emetic reflex arc revealed by expression of the immediate-early gene c-fos in the cat. J Neurosci 14:871–888

Miselis RR (1981) The efferent projections of the subfornical organ of the rat: a circumventricular organ within a neural network subserving water balance. Brain Res 230:1–23

Miselis RR, Shapiro RE, Hand PJ (1979) Subfornical organ efferents to neural systems for control of body water. Science 205:1022–1025

Miselis RR, Shapiro RR, Hyde TM (1987a) The area postrema. In: Gross PM (ed) Circumventricular Organs and Body Fluids. CRC Press, Boca Raton, pp 185–207 vol II

Miselis RR, Weiss ML, Shapiro RE (1987b) Modulation of the visceral neuraxis. In: Gross PM (ed) Circumventricular Organs and Body Fluids. CRC Press, Boca Raton, pp 143–162 vol III

Moellenhoff E, Lebrun CJ, Blume A, Culman J, Herdegen T, Unger T (1998) Central angiotensin AT1 and muscarinic receptors in ITF expression on intracerebroventricular NaCl. Am J Physiol 275:R234-R244

Morgan JI, Cohen DR, Hempstead JL, Curran T (1987) Mapping patterns of c-fos expression in the central nervous system after seizure. Science 237:192–197

Morgan JI, Curran T (1989a) Calcium and proto-oncogene involvement in the immediate-early response in the nervous system. Ann N Y Acad Sci 568:283–290

Morgan JI, Curran T (1989b) Stimulus-transcription coupling in neurons: role of cellular immediate- early genes. Trends Neurosci 12:459–462

Morien A, Garrard L, Rowland NE (1999) Expression of Fos immunoreactivity in rat brain during dehydration: effect of duration and timing of water deprivation. Brain Res 816:1–7

Mufson EJ, Cai WJ, Jaffar S, Chen E, Stebbins G, Sendera T, Kordower JH (1999) Estrogen receptor immunoreactivity within subregions of the rat forebrain: neuronal distribution and association with perikarya containing choline acetyltransferase. Brain Res 849:253–274

Mugnaini E, Oertel WH (1985) An atlas of the distribution of GABAergic neurons and terminals in the rat CNS as revealed by GAD immunohistochemistry. In: Bjorklund A, Hokfelt T (eds) Handbook of Chemical Neuroanatomy. Vol. 4. GABA and Neuropeptides in the CNS. Part I. Elsevier, Amsterdam, pp 436–608

Mumford AD, Parry LJ, Summerlee AJ (1989) Lesion of the subfornical organ affects the haemotensive response to centrally administered relaxin in anaesthetized rats. J Endocrinol 122:747–755

Murone C, Paxinos G, McKinley MJ, Oldfield BJ, Muller-Esterl W, Mendelsohn FA, Chai SY (1997) Distribution of bradykinin B2 receptors in sheep brain and spinal cord visualized by in vitro autoradiography. J Comp Neurol 381:203–218

Nadeau S, Rivest S (1999) Regulation of the gene encoding tumor necrosis factor alpha (TNF-alpha) in the rat brain and pituitary in response in different models of systemic immune challenge. J Neuropathol Exp Neurol 58:61–77

Nadeau S, Rivest S (2000) Role of microglial-derived tumor necrosis factor in mediating CD14 transcription and nuclear factor kappa B activity in the brain during endotoxemia. J Neurosci 20:3456–3468

Nakamori T, Morimoto A, Yamaguchi K, Watanabe T, Long NC, Murakami N (1993) Organum vasculosum laminae terminalis (OVLT) is a brain site to produce interleukin-1 beta during fever. Brain Res 618:155–159

Nakamura K, Kaneko T, Yamashita Y, Hasegawa H, Katoh H, Ichikawa A, Negishi M (1999) Immunocytochemical localization of prostaglandin EP3 receptor in the rat hypothalamus. Neurosci Lett 260:117–120

Nambu T, Sakurai T, Mizukami K, Hosoya Y, Yanagisawa M, Goto K (1999) Distribution of orexin neurons in the adult rat brain. Brain Res 827:243–260

Nazarali AJ, Gutkind JS, Correa FM, Saavedra JM (1990) Decreased angiotensin II receptors in subfornical organ of spontaneously hypertensive rats after chronic antihypertensive treatment with enalapril. Am J Hypertens 3:59–61
Nelson DO (1988) Altered angiotensin II sensitivity of neurons in the organum vasculosum lamina terminalis region of the spontaneously hypertensive rat. Brain Res 444:46–52
Newton BW, Maley B (1985a) Localization of somatostatin-like immunoreactivity in the area postrema of the rat and cat. Neurosci Lett 54:333–338
Newton BW, Maley B, Traurig H (1985) The distribution of substance P, enkephalin, and serotonin immunoreactivities in the area postrema of the rat and cat. J Comp Neurol 234:87–104
Newton BW, Maley BE (1985b) Cholecystokinin-octapeptide like immunoreactivity in the area postrema of the rat and cat. Regul Pept 13:31–40
Newton BW, Maley BE (1985c) Distribution of neurotensin-like immunoreactivity in rat and cat area postrema. Peptides 6:301–306
Newton BW, Maley BE (1987) A comparison of GABA- and GAD-like immunoreactivity within the area postrema of the rat and cat. J Comp Neurol 255:208–216
Nissen R, Bourque CW, Renaud LP (1993) Membrane properties of organum vasculosum lamina terminalis neurons recorded in vitro. Am J Physiol 264:R811-R815
Niwa M, Shigematsu K, Kurihara M, Kataoka Y, Maeda T, Nakao K, Imura H, Matsuo H, Tsuchiyama H, Ozaki M (1988) Receptor autoradiographic evidence of specific brain natriuretic peptide binding sites in the porcine subfornical organ. Neurosci Lett 95:113–118
O'Donnell D, Ahmad S, Wahlestedt C, Walker P (1999) Expression of the novel galanin receptor subtype GALR2 in the adult rat CNS: distinct distribution from GALR1. J Comp Neurol 409:469–481
Oldendorf WH (1971) Brain uptake of radiolabeled amino acids, amines, and hexoses after arterial injection. Am J Physiol 221:1629–1639
Oldfield B, Hards DK, McKinley MJ (1991) Projections from the subfornical organ to the supraoptic nucleus in the rat: ultrastructural identification of an interposed synapse in the median preoptic nucleus using a combination of neural tracers. Brain Res 558:13–19
Oldfield BJ, Badoer E, Hards DK, McKinley MJ (1994) Fos production in retrogradely labelled neurons of the lamina terminalis following intravenous infusion of either hypertonic saline or angiotensin II. Neuroscience 60:255–262
Oldfield BJ, Bicknell RJ, McAllen RM, Weisinger RS, McKinley MJ (1991) Intravenous hypertonic saline induces Fos immunoreactivity in neurons throughout the lamina terminalis. Brain Res 561:151–156
Oldfield BJ, Ganten D, McKinley MJ (1989) An ultrastructural analysis of the distribution of angiotensin II in the rat brain. J Neuroendocrinol 1:121–128
Oldfield BJ, Hards DK, McKinley MJ (1992) Neurons in the median preoptic nucleus of the rat with collateral branches to the subfornical organ and supraoptic nucleus. Brain Res 586:86–90
Oldfield BJ, McKinley MJ (1994b) Distribution of Fos in rat brain resulting from endogenously-generated angiotensin II. Kidney Int 46:1567–1569
Oldfield BJ, Miselis RR, McKinley MJ (1991) Median preoptic nucleus projections to vasopressin-containing neurons of the supraoptic nucleus in sheep—a light and electron microscope study. Brain Res 542:193–200
Olsson K (1972) Dipsogenic effects of intracarotid infusions of various hyperosmolal solutions. Acta Physiol Scand 85:517–522
Osheroff PL, Phillips HS (1991) Autoradiographic localization of relaxin binding sites in rat brain. Proc Natl Acad Sci U S A 88:6413–6417
Otsuka A, Barnes KL, Ferrario CM (1986) Contribution of area postrema to pressor actions of angiotensin II in dog. Am J Physiol 251:H538-H546
Palkovits M (1987) Summary of structural and functional aspects of the circumventricular organs. In: Gross PM (ed) Circumventricular Organs and Body Fluids. CRC Press, Boca Raton, pp 209–218 vol II
Palkovits M, Brownstein MJ (1985) Distribution of neuropeptides in the central nervous system using biochemical micromethods. In: Bjorklund A, Hokfelt T (eds) Handbook of Chemical

Neuroanatomy. Vol. 4. GABA and Neuropeptides in the CNS. Part I. Elsevier, Amsterdam, pp 1–71
Palkovits M, Mezey E, Ambach G, Kivovics P (1978) Neural and vascular connections between the prganum vasculosum laminae terminalis and preoptic nuclei. In: Brain Endocrine Interaction III. Neural hormones and Reproduction. Karger, Basel, pp 302–312
Papas S, Smith P, Ferguson AV (1990) Electrophysiological evidence that systemic angiotensin influences rat area postrema neurons. Am J Physiol 258:R70-R76
Park RG, Clevers J, McKinley MJ, Rundgren M. (1989) Renal denervation does not prevent dehydration induced natriuresis in sheep. Acta Physiol Scand 137:199–206
Parry LJ, Poterski RS, Summerlee AJ (1994) Effects of relaxin on blood pressure and the release of vasopressin and oxytocin in anesthetized rats during pregnancy and lactation. Biol Reprod 50:622–628
Parry LJ, Poterski RS, Summerlee AJS, Jones SA (1990) Mechanism of the haemotensive action of porcine relaxin in anaesthetised rats. J Neuroendocrinol 2:53–58
Parry LJ, Summerlee AJ (1991) Central angiotensin partially mediates the pressor action of relaxin in anesthetized rats. Endocrinology 129:47–52
Pastuskovas C, Vivas L (1997) Effect of intravenous captopril on c-fos expression induced by sodium depletion in neurons of the lamina terminalis. Brain Res Bull 44:233–236
Patronas P, Horowitz M, Simon E, Gerstberger R (1998) Differential stimulation of c-fos expression in hypothalamic nuclei of the rat brain during short-term heat acclimation and mild dehydration. Brain Res 798:127–139
Peck JW, Blass EM (1975) Localization of thirst and antidiuretic osmoreceptors by intracranial injections in rats. Am J Physiol 228:1501–1509
Petrusz P, Merchenthaler I, Maderdrut JL (1985) Distribution of enkephalin-containing neurons in the central nervous system. In: Bjorklund A, Hokfelt T (eds) Handbook of Chemical Neuroanatomy. Vol. 4. GABA and Neuropeptides in the CNS. Part I. Elsevier, Amsterdam, pp 273–334
Pfenninger K (1969) Subfornikalorgan und Liquor cerebrospinalis. In: Sterba G (ed) Zirkumventrikulare Organe und Liquor. VEB Gustav Fischer Verlag, Jena, pp 103–106
Phillips MI, Balhorn L, Leavitt M, Hoffman W (1974) Scanning electron microscope study of the rat subfornical organ. Brain Res 80:95–110
Phillips MI, Camacho A (1987) Neural connections of the organum vasculosum of the lamina terminalis. In: Gross PM (ed) Circumventricular Organs and Body Fluids. CRC Press, Boca Raton, pp 157–169 vol I
Phillips MI, Felix D (1976) Specific angiotensin II receptive neurons in the cat subfornical organ. Brain Res 109:531–540
Phillips PA, Abrahams JM, Kelly J, Paxinos G, Grzonka Z, Mendelsohn FA, Johnston CI (1988) Localization of vasopressin binding sites in rat brain by in vitro autoradiography using a radioiodinated V1 receptor antagonist. Neuroscience 27:749–761
Pickel VM (1985) General morphological features of peptidergic neurons. In: Bjorklund A, Hokfelt T (eds) Bjorklund A, Hokfelt T (eds) Handbook of Chemical Neuroanatomy. Vol. 4. GABA and Neuropeptides in the CNS. Part I. Elsevier, Amsterdam, pp 72–92
Pickel VM, Armstrong DM (1984) Ultrastructural localization of monoamines and peptides in rat area postrema. Fed Proc 43:2949–2951
Pickel VM, Chan J, Ganten D (1986) Dual peroxidase and colloidal gold-labeling study of angiotensin converting enzyme and angiotensin-like immunoreactivity in the rat subfornical organ. J Neurosci 6:2457–2469
Pines L (1926) Über ein unbeachteles gebille im Gehirn einiger Säugetiere das subfornicale Organe des 3. Ventricles. J Psychol Neurol 34:36–57
Piva F, Limonta P, Martini L (1982) Role of the organum vasculosum laminae terminalis in the control of gonadotrophin secretion in rats. J Endocrinol 93:355–364
Polson JW, Potts PD, Li YW, Dampney RA (1995) Fos expression in neurons projecting to the pressor region in the rostral ventrolateral medulla after sustained hypertension in conscious rabbits. Neuroscience 67:107–123
Potts PD, Hirooka Y, Dampney RA (1999) Activation of brain neurons by circulating angiotensin II: direct effects and baroreceptor-mediated secondary effects. Neuroscience 90:581–94

Potts PD, Ludbrook J, Gillman-Gaspari TA, Horiuchi J, Dampney RA (2000) Activation of brain neurons following central hypervolaemia and hypovolaemia: contribution of baroreceptor and non-baroreceptor inputs. Neuroscience 95:499–511
Potts PD, Polson JW, Hirooka Y, Dampney RAL (1997) Effects of sinoaortic denervation on Fos expression in the brain evoked by hypertension and hypotension in conscious rabbits. Neurosci 77:503–520
Putnam TJ (1922) The intercolumnar tubercle, an undescribed area in the anterior wall of the third ventricle. Bull Johns Hopk Hosp 33:181–182
Quan N, Stern EL, Whiteside MB, Herkenham M (1999) Induction of pro-inflammatory cytokine mRNAs in the brain after peripheral injection of subseptic doses of lipopolysaccharide in the rat. J Neuroimmunol 93:72–80
Quan N, Whiteside M, Herkenham M (1998) Time course and localization patterns of interleukin-1beta messenger RNA expression in brain and pituitary after peripheral administration of lipopolysaccharide. Neuroscience 83:281–293
Quan N, Whiteside M, Kim L, Herkenham M (1997) Induction of inhibitory factor kappaBalpha mRNA in the central nervous system after peripheral lipopolysaccharide administration: an in situ hybridization histochemistry study in the rat. Proc Natl Acad Sci U S A 94:10985–10990
Quinn SJ, Kifor O, Trivedi S, Diaz R, Vassilev P, Brown E (1998) Sodium and ionic strength sensing by the calcium receptor. J Biol Chem 273:19579–19586
Quirion R, Boksa P (1986) Autoradiographic distribution of muscarinic [3H]acetylcholine receptors in rat brain: comparison with antagonists. Eur J Pharmacol 123:170–172
Quirion R, Dalpe M, Dam TV (1986) Characterization and distribution of receptors for the atrial natriuretic peptides in mammalian brain. Proc Natl Acad Sci U S A 83:174–178
Quirion R, Martel J-C (1992) Brain neuropeptide Y receptors: distribution and possible relevance to function. In: Bjorklund A, Hokfelt T, Kumar MJ (eds) Handbook of Chemical Neuroanatomy. Neuropeptide receptors in the CNS. Elsevier, Amsterdam, pp 247–287
Ramsay DJ, Keil LC, Sharpe MC, Shinsaki J (1978) Angiotensin II infusion increases vasopressin, ACTH and 11-hydroxycorticosteroid secretion. Am. J. Physiol. 234:R66-R71
Ray PE, Castren E, Ruley EJ, Saavedra JM (1990a) Different effects of sodium or chloride depletion on angiotensin II receptors in rats. Am J Physiol 258:R1008-R1015
Ray PE, Ruley EJ, Saavedra JM (1990b) Down-regulation of angiotensin II receptors in subfornical organ of young male rats by chronic dietary sodium depletion. Brain Res 510:303–308
Reid IA (1984) Actions of angiotensin II on the brain: mechanisms and physiologic role. Am J Physiol 246:F533–43
Renaud LP, Cunningham JT, Nissen R, Yang CR (1993) Electrophysiology of central pathways controlling release of neurohypophysial hormones. Focus on the lamina terminalis and diagonal band inputs to the supraoptic nucleus. Ann N Y Acad Sci 689:122–32
Renaud LP, Ferguson AV, Day TA, Bourque CW, Sgro S (1985) Electrophysiology of the subfornical organ and its hypothalamic connections–an in-vivo study in the rat. Brain Res Bull 15:83–86
Reynolds DJ, Lowenstein PR, Moorman JM, Grahame-Smith DG, Leslie RA (1994) Evidence for cholinergic vagal afferents and vagal presynaptic M1 receptors in the ferret. Neurochem Int 25:455–464
Reynolds DJM, Barber NA, Grahame-Smith DG, Leslie RA (1991) Cisplatin-evoked induction of c-fos protein in the brainstem of the ferret: the effect of cervical vagotomy and the anti-emetic 5-HT3 receptor antagonist granisetron (BRL 43694). Brain Res 565:231–236
Ribeiro-da-Silva A, McLeod AL, Krause JE (2000) Neurokinin receptors in the CNS. In: Quirion R, Bjorklund A, Hokfelt T (eds) Handbook of Chemical Neuroanatomy. Peptide Receptors part I. Elsevier, Amsterdam, pp 195–240
Richard D, Bourque CW (1992) Synaptic activation of rat supraoptic neurons by osmotic stimulation of the organum vasculosum lamina terminalis. Neuroendocrinology 55:609–611
Richard D, Bourque CW (1995) Synaptic control of rat supraoptic neurones during osmotic stimulation of the organum vasculosum lamina terminalis in vitro. J Physiol 489:567–577
Riediger T, Rauch M, Schmid HA (1999) Actions of amylin on subfornical organ neurons and on drinking behavior in rats. Am J Physiol 276:R514-R521

Rinaman L, Stricker EM, Hoffman GE, Verbalis JG (1997) Central c-Fos expression in neonatal and adult rats after subcutaneous injection of hypertonic saline. Neuroscience 79:1165–1175
Rittenhouse PA, Skoog KM, Blair ML, Sladek CD, Mangiapane ML (1986) mechanism of the anti-hypertensive effect of area postrema ablation in SHR. Fed Proc 45:875
Rogers KV, Dunn CK, Hebert SC, Brown EM (1997) Localization of calcium receptor mRNA in the adult rat central nervous system by in situ hybridization. Brain Res 744:47–56
Rogerson FM, Schlawe I, Paxinos G, Chai SY, McKinley MJ, Mendelsohn FA (1995) Localization of angiotensin converting enzyme by in vitro autoradiography in the rabbit brain. J Chem Neuroanat 8:227–243
Rohlich P, Wenger T (1969) Elektronenmikroscopische Untersuchungen am organon vasculosum laminae terminalis der Ratte. Z. Zellforsch. 102:483–506
Rohr VU (1966) Zum Feinbau des Subfornikalorgans der Katze. II. Neurosekretorische Aktivitat. Z. Zellforsch. 75:11–34
Rohrschneider I, Schinko I, Wetzstein R (1972) Der Feinbau der Are postrema der Maus. Z Zellforsch Mikrosk Anat 123:251–276
Rosas-Arellano MP, Solano-Flores LP, Ciriello J (1996a) Arcuate nucleus inputs onto subfornical organ neurons that respond to plasma hypernatremia and angiotensin II. Brain Res 707:308–313
Rosas-Arellano MP, Solano-Flores LP, Ciriello J (1996b) Neurotensin projections to subfornical organ from arcuate nucleus. Brain Res 706:323–327
Roth GI, Yamamoto WS (1968) The microcirculation of the area postrema in the rat. J. Comp. Neurol. 133:329–340
Rowland NE, Crews EC, Gentry RM (1997) Comparison of Fos induced in rat brain by GLP-1 and amylin. Regul Pept 71:171–174
Rowland NE, Fregly MJ, Han L, Smith G (1996) Expression of Fos in rat brain in relation to sodium appetite: furosemide and cerebroventricular renin. Brain Res 728:90–96
Rowland NE, Li BH, Rozelle AK, Smith GC (1994) Comparison of fos-like immunoreactivity induced in rat brain by central injection of angiotensin II and carbachol. Am J Physiol 267:R792-R798
Rudert H, Schwink A, Wetzstein R (1968) Die Feinstruktur des Subfornikalorgans beim Kaninchen. II. Das neuronal und gliale gewebel. Z. Zellforsch. 88:145–179
Ruggiero DA, Regunathan S, Wang H, Milner TA, Reis DJ (1998) Immunocytochemical localization of an imidazoline receptor protein in the central nervous system. Brain Res 780:270–293
Rundgren M, Fyhrquist F (1978) A study of permanent adipsia induced by medial forebrain lesions. Acta Physiol Scand 103:463–471
Saavedra JM, Chevillard C (1982) Angiotensin-converting enzyme is present in the subfornical organ and other circumventricular organs of the rat. Neurosci Lett 29:123–127
Saavedra JM, Correa FM, Kurihara M, Shigematsu K (1986) Increased number of angiotensin II receptors in the subfornical organ of spontaneously hypertensive rats. J Hypertens Suppl 4:S27-S30
Saavedra JM, De Oliveira AM, Johren), Tonelli L (2000) Brain endothelin and natriuretic peptide receptors. In: Quirion R, Björklund A, Hökfelt T (eds) Handbook of Chemical Neuroanatomy Volume 16. Peptide Receptors, Part 1. Elsevier Science, Amsterdam pp 125–162
Saavedra JM, Israel A, Kurihara M (1987) Increased atrial natriuretic peptide binding sites in the rat subfornical organ after water deprivation. Endocrinology 120:426–428
Saavedra JM, Zorad S, Tsutsumi K (1992) Localization of atrial natriuretic peptide B and C receptors in rat brain. In: Bjorklund A, Hokfelt T, Kumar MJ (eds) Handbook of Chemical Neuroanatomy. Neuropeptide receptors in the CNS. Elsevier, Amsterdam, pp 39–54
Sagar SM, Price KJ, Kasting NW, Sharp FR (1995) Anatomic patterns of Fos immunostaining in rat brain following systemic endotoxin administration. Brain Res Bull 36:381–392
Samson WK, Aguila MC, Martinovic J, Antunes-Rodrigues J, Norris M (1987) Hypothalamic action of atrial natriuretic factor to inhibit vasopressin secretion. Peptides 8:449–454
Samson WK, Snyder G, Fawcett CP, McCann SM (1980) Chromatographic and biologic analysis of ME and OVLT LHRH

Saper CB, Levisohn D (1983) Afferent connections of the median preoptic nucleus in the rat: anatomical evidence for a cardiovascular integrative mechanism in the anteroventral third ventricular (AV3V) region. Brain Res 288:21–31
Sayer RJ, Hubbard JI, Sirett NE (1984) Rat organum vasculosum laminae terminalis in vitro: responses to transmitters. Am J Physiol 247:R374-R379
Schinko I, Rohrschneider I, Wetzstein R (1972) Electronmikroscopische Untersuchengen am Subfornikal der Maus. Z Zellforsch Mikrosk Anat 123:277–294
Schmid HA (1995) Regional differences in the blood-brain barrier of the subfornical organ of rats and ducks *(Anas platyrhyncos)*. J Brain Res 36:565–574
Schmid HA (1998) Effect of glutamate and angiotensin II on whole cell currents and release of nitric oxide in the rat subfornical organ. Amino Acids 14:113–119
Schmid HA, Simon E (1992) Effect of angiotensin II and atrial natriuretic factor on neurons in the subfornical organ of ducks and rats in vitro. Brain Res 588:324–328
Schoorlemmer GH, Johnson AK, Thunhorst RL (2000) Effect of hyperosmotic solutions on salt excretion and thirst in rats. Am J Physiol Regul Integr Comp Physiol 278:R917–923
Schreihofer DA, Cameron JL, Verbalis JG, Rinaman L (1997) Cholecystokinin induces Fos expression in catecholaminergic neurons of the macaque monkey caudal medulla. Brain Res 770:37–44
Schwendemann G (1973) Zur Ultrastruktur des Organon vsaculosum laminae terminalis der Ratte mit besonderer Berucksichtigung der Gefaße. Advances in Anatomy, Embryology and Cell Biology 47:7–72
Scott CJ, Tilbrook AJ, Simmons DM, Rawson JA, Chu S, Fuller PJ, Ing NH, Clarke IJ (2000) The distribution of cells containing estrogen receptor-alpha (ER alpha) and ER beta messenger ribonucleic acid in the preoptic area and hypothalamus of the sheep: comparison of males and females. Endocrinology 141:2951–2962
Scott DE, Paull WK (1983) Scanning electron microscopy of the mammalian cerebral-ventricular system. Micron 14:165–186
Scrogin KE, Johnson AK, Schmid HA (1998) Multiple receptor subtypes mediate the effects of serotonin on rat subfornical organ neurons. Am J Physiol 275:R2035-R2042
Sexton PM, Albiston A, Morfis M, Tilakaratne N (2001) Receptor activity modifying proteins. Cell Signal 13:73–83
Sexton PM, McKenzie JS, Mason RT, Moseley JM, Martin TJ, Mendelsohn FA (1986) Localization of binding sites for calcitonin gene-related peptide in rat brain by in vitro autoradiography. Neuroscience 19:1235–1245
Sexton PM, Paxinos G, Huang XF, Mendelsohn FA (1994a) In vitro autoradiographic localization of calcitonin binding sites in human medulla oblongata. J Comp Neurol 341:449–463
Sexton PM, Paxinos G, Kenney MA, Wookey PJ, Beaumont K (1994b) In vitro autoradiographic localization of amylin binding sites in rat brain. Neuroscience 62:553–567
Shapiro RE, Miselis RR (1985) The central neural connections of the area postrema of the rat. J Comp Neurol 234:344–364
Shaver SW, Sposito NM, Gross PM (1990) Quantitative fine structure of capillaries in subregions of the rat subfornical organ. J Comp Neurol 294:145–152
Sibbald JR, Hubbard JI, Sirett NE (1988) Responses from osmosensitive neurons of the rat subfornical organ in vitro. Brain Res 461:205–214
Sibbald JR, Hubbard JI, Sirett NE (1988) Responses from osmosensitive neurons of the rat subfornical organ in vitro. Brain Res 461:205–214
Simpson JB, Epstein AN, Camardo JS, Jr. (1978) Localization of receptors for the dipsogenic action of angiotensin II in the subfornical organ of rat. J Comp Physiol Psychol 92:581–601
Simpson JB, Routtenberg A (1973) Subfornical organ: site of drinking elicitation by angiotensin II. Science 181:1772–1775
Sinnayah P, Burns P, Wade JD, Weisinger RS, McKinley MJ (1999) Water drinking in rats resulting from intravenous relaxin and its modification by other dipsogenic factors. Endocrinology 140:5082–5086
Skofitsch G, Sills MA, Jacobowitz DM (1986) Autoradiographic distribution of 125I-galanin binding sites in the rat central nervous system. Peptides 7:1029–1042

Sladek CD, Johnson AK (1983) Effect of anteroventral third ventricle lesions on vasopressin release by organ-cultured hypothalamo-neurohypophyseal explants. Neuroendocrinology 37:78–84
Sly DJ, Colvill L, McKinley MJ, Oldfield BJ (1999) Identification of neural projections from the forebrain to the kidney, using the virus pseudorabies. J Auton Nerv Syst 77:73–82
Smith GE (1898) Further observations on the fornix, with special reference to the brain of Nictophilus. J Anat Physiol (Lond) 32:231–246
Sofroniew MV (1985) Vasopressin- and neurophysin-immunoreactive neurons in the septal region, medial amygdala and locus coeruleus in colchicine-treated rats. Neuroscience 15:347–358
Solano-Flores LP, Rosas-Arellano MP, Ciriello J (1997) Fos induction in central structures after afferent renal nerve stimulation. Brain Res 753:102–119
Song K, Allen AM, Paxinos G, Mendelsohn FA (1992) Mapping of angiotensin II receptor subtype heterogeneity in rat brain. J Comp Neurol 316:467–484
Speth RC, Wamsley JK, Gehlert DR, Chernicky CL, Barnes KL, Ferrario CM (1985) Angiotensin II receptor localization in the canine CNS. Brain Res 326:137–143
Spoerri O (1963) Über die Gefässversorgung des Sufornikalorganes der Ratte. Acta Anat (Basel) 54:333–348
Sposito NM, Gross PM (1987) Topography and morphometry of capillaries in the rat subfornical organ. J Comp Neurol 260:36–46
Steardo L, Steardo MD, Testa N, Attanasio A, Persichella M, Cuomo V (1994) Atrial natriuretic factor antagonises angiotensin II-induced vasopressin release in rat subfornical organ. Acta Neurol (Napoli) 16:229–234
Stitt JT (1985) Evidence for the involvement of the organum vasculosum laminae terminalis in the febrile response of rabbits and rats. J Physiol 368:501–511
Stitt JT (1986) Prostaglandins as the neural mediator of the febrile response. Yale J Biol Med 59:137–149
Stitt JT (1991) Differential sensitivity in the sites of fever production by prostaglandin E1 within the hypothalamus of the rat. J Physiol 432:99–110
Stitt JT, Shimada SG (1989) Enhancement of the febrile responses of rats to endogenous pyrogen occurs within the OVLT region. J Appl Physiol 67:1740–1746
Stornetta RL, Hawelu-Johnson CL, Guyenet PG, Lynch KR (1988) Astrocytes synthesize angiotensinogen in brain. Science 242:1444–1446
Stricker EM, Craver CF, Curtis KS, Peacock-Kinzig KA, Sved AF, Smith JC (2001) Osmoregulation in water-deprived rats drinking hypertonic saline: effect of area postrema lesions. Am J Physiol 280:R831-R842
Strominger NL, Knox AP, Carpenter DO (1994) The connectivity of the area postrema in the ferret. Brain Res Bull 33:33–47
Stumpf WE, Bidmon HJ, Ruhle HJ (1992) Steroid hormones and circumventricular organs. Prog Brain Res 91:271–277
Summerlee AJS, Robertson GF (1995) Central administration of porcine relaxin stimulates drinking behaviour in rats: an effect mediated by central angiotensin II. Endocrine 3:377–381
Summy-Long JY (1987) Biochemistry. In: Gross PM (ed) Circumventricular Organs and Body Fluids. CRC Press, Boca Raton, pp 59–78 vol I
Sun K, Ferguson AV (1996) Angiotensin II and glutamate influence area postrema neurons in rat brain slices. Regul Pept 63:91–98
Sunn N, Egli M, Burazin TCD, Burns P, Colvill L, Davern P, Denton DA, Oldfield BJ, Weisinger RS, Rauch M, Schmid HA, McKinley MJ (2002) Circulating relaxin acts on subfornical organ neurons to stimulate water drinking in the rat. Proc Soc Natl Acad Sci USA 99:1701–1706
Sunn N, McKinley MJ, Oldfield BJ (2001) Identification of efferent neural pathways from the lamina terminalis activated by blood-borne relaxin. J Neuroendocrinol 13:432–437
Sunn N, McKinley MJ, Oldfield BJ (2003) Circulating angiotensin II activates neurons in circumventricular organs that project to the bed nucleus of the stria terminalis. J Neuroendocrinol (In Press)
Swaminathan S (1980) Osmoreceptors or sodium receptors: an investigation into ADH release in the rhesus monkey. J Physiol (Lond) 307:71–83

Swanson LW, Lind RW (1986) Neural projections subserving the initiation of a specific motivated behavior in the rat: new projections from the subfornical organ. Brain Res 379:399–403
Takahashi Y, Smith P, Ferguson A, Pittman QJ (1997) Circumventricular organs and fever. Am J Physiol 273:R1690-R1695
Takeuchi Y, Sano Y (1983) Serotonin distribution in the circumventricular organs of the rat. An immunohistochemical study. Anat Embryol 167:311–319
Takumi T, Taguchi K, Miyake S, Sakakida Y, Takashima N, Matsubara C, Maebayashi Y, Okumura K, Takekida S, Yamamoto S, Yagita K, Yan L, Young MW, Okamura H (1988) A light-independent oscillatory gene *mPer3* in mouse SCN and OVLT. EMBO J 17:4753–4759
Tan YY, Wade JD, Tregear GW, Summers RJ (1999) Quantitative autoradiographic studies of relaxin binding in rat atria, uterus and cerebral cortex: characterization and effects of oestrogen treatment. Br J Pharmacol 127:91–98
Tanaka J, Hayashi Y, Shimamune S, Hori K, Nomura M (1997) Subfornical organ efferents enhance extracellular noradrenaline concentrations in the median preoptic area in rats. Neurosci Lett 230:171–174
Tanaka J, Hayashi Y, Shimamune S, Nomura M (1997) Ascending pathways from the nucleus of the solitary tract to the subfornical organ in the rat. Brain Res 777:237–241
Tanaka J, Hayashi Y, Watai T, Shimamune S (1997) Angiotensinergic modulation of osmotic activation of neurosecretory neurons. Neuroreport 8:2903–2906
Tanaka J, Kaba H, Saito H, Seto K (1985) Electrophysiological evidence that circulating angiotensin II sensitive neurons in the subfornical organ alter the activity of hypothalamic paraventricular neurohypophyseal neurons in the rat. Brain Res 342:361–365
Tanaka J, Kaba H, Saito H, Seto K (1985) Subfornical organ neurons with efferent projections to the hypothalamic paraventricular nucleus: an electrophysiological study in the rat. Brain Res 346:151–154
Tanaka J, Saito H, Kaba H (1987) Subfornical organ and hypothalamic paraventricular nucleus connections with median preoptic nucleus neurons: an electrophysiological study in the rat. Exp Brain Res 68:579–585
Tang-Christensen M, Larsen PJ, Goke R, Fink-Jensen A, Jessop DS, Moller M, Sheikh SP (1996) Central administration of GLP-1-(7–36) amide inhibits food and water intake in rats. Am J Physiol 271:R848-R856
Thornton SM, Fitzsimons JT (1995) The effects of centrally administered porcine relaxin on drinking behaviour in male and female rats. J Neuroendocrinol 7:165–169
Thrasher TN (1985) Circumventricular organs, thirst, and vasopressin secretion. In Schrier RW (ed): Vasopressin. Raven Press, New York. pp. 311–318
Thrasher TN, Brown CJ, Keil LC, Ramsay DJ (1980) Thirst and vasopressin release in the dog: an osmoreceptor or sodium receptor mechanism? Am J Physiol 238:R333-R339
Thrasher TN, Keil LC (1987) Regulation of drinking and vasopressin secretion: role of organum vasculosum laminae terminalis. Am J Physiol 253:R108-R120
Thrasher TN, Keil LC, Ramsay DJ (1982) Lesions of the organum vasculosum of the lamina terminalis (OVLT) attenuate osmotically-induced drinking and vasopressin secretion in the dog. Endocrinology 110:1837–1839
Thrasher TN, Simpson JB, Ramsay DJ (1982) Lesions of the subfornical organ block angiotensin-induced drinking in the dog. Neuroendocrinology 35:68–72
Thunhorst RL, Ehrlich KJ, Simpson JB (1990) Subfornical organ participates in salt appetite. Behav Neurosci 104:637–642
Thunhorst RL, Fitts DA (1994) Peripheral angiotensin causes salt appetite in rats. Am J Physiol 267:R171–177
Thunhorst RL, Fitts DA, Simpson JB (1989) Angiotensin-converting enzyme in subfornical organ mediates captopril- induced drinking. Behav Neurosci 103:1302–1310
Thunhorst RL, Xu Z, Cicha MZ, Zardetto-Smith AM, Johnson AK (1998) Fos expression in rat brain during depletion-induced thirst and salt appetite. Am J Physiol 274:R1807–1814
Torack RM, Finke EH (1971) Evidence for a sequestration of function within the area postrema based on scanning electron microscopy and the penetration of horseradish peroxidase. Z Zellforsch Mikrosk Anat 118:85–96

Tramu G, Pillez A, Leonardelli J (1983) Serotonin axons of the ependyma and circumventricular organs in the forebrain of the guinea pig. An immunohistochemical study. Cell Tissue Res 228:297–311
Tribollet E, Dreifuss JJ (1981) Localization of neurones projecting to the hypothalamic paraventricular nucleus area of the rat: a horseradish peroxidase study. Neuroscience 6:1315–28
Ugrumov MV, Taxi J, Mitskevich MS, Tramu G (1986) Development of the hypothalamic serotoninergic system during ontogenesis in rats. Immunocytochemical and radioautographic study. Brain Res 395:75–84
Undesser K P, Hasser E, Haywood JR, Johnson A K, Bishop VS (1985). Interactions of vasopressin with the area postrema in arterial baroreflex function in conscious rabbits. Circ Res 56: 410-417
Uschakov A, McAllen RM, Oldfield BJ, McKinley MJ (2001). Efferent projections of subpopulations of neurons in the lamina terminalis. Soc Neurosci Abstracts 27:733–738
Vallieres L, Rivest S (1997) Regulation of the gene encoding interleukin-6, its receptor, and gp130 in the rat brain in response to the immune activator lipopolysacharide and the proinflammatory cytokine interleukin-1beta. J Neurochem 69:1668–1683
van der Kooy D, Koda LY (1983) Organization of the projections of a circumventricular organ: the area postrema in the rat. J Comp Neurol 219:328–338
Van Houten M, Mangiapane ML, Reid IA, Ganong WF (1983) [Sar,Ala]angiotensin II in cerebrospinal fluid blocks the binding of blood-borne [125I]angiotensin II to the circumventricular organs. Neuroscience 10:1421–1426
Vanderhaeghen JJ (1985) Neuronal cholecystokinin. In: Bjorklund A, Hokfelt T (eds) Handbook of Chemical Neuroanatomy. Vol. 4. GABA and Neuropeptides in the CNS. Part I. Elsevier, Amsterdam, pp 406–435
Verney EB (1947) The antidiuretic hormone and factors which determine its release. Proc Roy Soc B 135:25–106
Vivas L, Chiaraviglio E, Carrer HF (1990) Rat organum vasculosum laminae terminalis in vitro: responses to changes in sodium concentration. Brain Res 519:294–300
Vivas L, Pastuskovas CV, Tonelli L (1995) Sodium depletion induces Fos immunoreactivity in circumventricular organs of the lamina terminalis. Brain Res 679:34–41
Voisin DL, Simonian SX, Herbison AE (1997) Identification of estrogen receptor-containing neurons projecting to the rat supraoptic nucleus. Neuroscience 78:215–228
Wada E, McKinnon D, Heinemann S, Patrick J, Swanson LW (1990) The distribution of mRNA encoded by a new member of the neuronal nicotinic acetylcholine receptor gene family (alpha 5) in the rat central nervous system. Brain Res 526:45–53
Walberg F, Ottersen OP (1992) Neuroactive amino acids in the area postrema. An immunocytochemical investigation in rat with some observations in cat and monkey (Macaca fascicularis). Anat Embryol 185:529–545
Walker N, Lepee-Lorgeoux I, Fournier J, Betancur C, Rostene W, Ferrara P, Caput D (1998) Tissue distribution and cellular localization of the levocabastine- sensitive neurotensin receptor mRNA in adult rat brain. Brain Res Mol Brain Res 57:193–200
Wang L, Martinez V, Barrachina MD, Tache Y (1998) Fos expression in the brain induced by peripheral injection of CCK or leptin plus CCK in fasted lean mice. Brain Res 791:157–166
Washburn DL, Beedle AM, Ferguson AV (1999) Inhibition of subfornical organ neuronal potassium channels by vasopressin. Neuroscience 93:349–359
Washburn DL, Smith PM, Ferguson AV (1999) Control of neuronal excitability by an ion-sensing receptor. Eur J Neurosci 11:1947–1954
Watanabe E, Fujikawa A, Matsunaga H, Yasoshima Y, Sako N, Yamamoto T, Saegusa C, Noda M (2000) Nav2/NaG channel is involved in control of salt-intake behavior in the CNS. J Neurosci 20:7743–7751
Weindl A (1973) Neuroendocrine aspects of cicumventricular organs. In: Ganong WF, Martini L (eds) Frontiers in Neuroendocrinology. Oxford University Press, London, pp 3–32
Weindl A, Bufler J, Winkler B, Arzberger T, Hatt H (1992) Neurotransmitters and receptors in the subfornical organ. Immunohistochemical and electrophysiological evidence. Prog Brain Res 91:261–269

Weindl A, Joynt RJ (1972) Ultrastructure of the ventricular walls. Three-dimensional study of regional specialization. Arch Neurol 26:420–427
Weindl A, Schwink A, Wetzstein R (1967) Der Feinbau des Gefäßorgans der Lamina terminalis beim Kaninchen. I. Die Gefäße. Z Zellforsch 79:1–48
Weindl A, Schwink A, Wetzstein R (1968) Der Feinbau des Gefäßorgans de Lamina terminalis beim Kaninchen. II. Das neuronal und gliale Gefäße. Z Zellforsch 85:552–600
Weindl A, Sofroniew M (1985) Neuroanatomical pathways related to vasopressin. In: Ganten D, Pfaff D (eds) Current Topics in Neuroendocrinology. Neurobiology of Vasopressin. Springer-Verlag, Berlin, pp 137–196
Weisinger RS, Blair-West JR, Burns P, Denton DA, McKinley MJ (1996) The role of angiotensin II in ingestive behaviour: a brief review of angiotensin II thirst and Na appetite. Reg Pept 661:73–81
Weisinger RS, Burns P, Colvill LM, Davern P, Giles ME, Oldfield BJ, McKinley MJ (2000) Fos immunoreactivity in the lamina terminalis of adrenalectomized rats and effects of angiotension II type 1 receptor blockade or deoxycorticosterone. Neuroscience 98:167–180
Weisinger RS, Burns P, Eddie LW, Wintour EM (1993) Relaxin alters the plasma osmolality-arginine vasopressin relationship in the rat. J Endocrinol 137:505–510
Weisinger RS, Denton DA, Di Nicolantonio R, Hards DK, McKinley MJ, Oldfield B, Osborne PG (1990) Subfornical organ lesion decreases sodium appetite in the sodium- depleted rat. Brain Res 526:23–30
Weisinger RS, Denton DA, Di Nicolantonio R, McKinley MJ, Muller AF, Tarjan E (1987) Role of angiotensin in sodium appetite of sodium-deplete sheep. Am. J Physiol 253:R482-R488
Weiss ML, Hatton GI (1990) Collateral input to the paraventricular and supraoptic nuclei in rat. I. Afferents from the subfornical organ and the anteroventral third ventricle region. Brain Res Bull 24:231–238
Wenger T, Kerdelhue B, Halasz B (1981) Does the organum vasculosum of the lamina terminalis play a role in the regulation of reproduction? Acta Physiol Acad Scient Hung 58:257–267
Wenger T, Leonardelli J (1980) circadian and cyclic LHRH variations in the organum vasculosum of the lamina terminalis of female and male rats. Neuroendocrinol 31:331–337
Wenger T, Toro I (1971) Studies on the organon vasculosum laminae terminalis. IV. Fine structure of the organon vasculosum laminae terminalis in man. Acta Biol 22:331–342
Westerhaus MJ, Loewy AD (1999) Sympathetic-related neurons in the preoptic region of the rat identified by viral transneuronal labeling. J Comp Neurol 414:361–378
Wilkin LD, Mitchell LD, Ganten D, Johnson AK (1989) The supraoptic nucleus: afferents from areas involved in control of body fluid homeostasis. Neuroscience 28:573–584
Wilson JT (1906a) On the anatomy of the calamus region in the human brain: with an account of a hitherto undescribed "nucleus postremus". Part I. J Anat Physiol 40:210–241
Wilson JT (1906b) On the anatomy of the calamus region in the human brain: with an account of a hitherto undescribed "nucleus postremus". Part II. J Anat Physiol 40:357–386
Wise BL, Ganong WF (1960) The effect of ablation of the area postrema on water and electrolyte metabolism in dogs. Acta Neuroveg 22:14–32
Wise BL, Goldfien A, Ganong WF (1960) Endocrine function in dogs after ablation of the area postrema. Acta Neuroveg 22:1–13
Wisliocki GB, Putnam TJ (1920) Note on the anatomy of the areae postremae. Anat Rec 19:281–285
Wislocki GB, Putnam TJ (1924) Further observations on the anatomy and physiology of the areae postremae. Anat Rec 27:151–156
Wislocki GB, King LS (1936) The permeability of the hypophysis and hypothalamus to vital dyes, with a study of the hypophyseal vascular supply. Amer J Anat 58:421–472
Wislocki GB, Leduc EH (1952) Vital staining of the hematoencephalic barrier by silver nitrate and trypan blue, and cytological comparisons of the neurohypophysis, pineal body, area postrema, intercolumnar tubercle and supraoptic crest. J Comp Neurol 96:371–414
Woods JW, Bard P, Bleier R (1966) Functional capacity of the deafferented hypothalamus: water balance and responses to osmotic stimuli in the decerebrate cat and rat. J Neurophysiol 29:751–767

Wright SD, Ramos RA, Tobias PS, Ulevitch RJ, Mathison JC (1990) CD14, a receptor for complexes of lipopolysaccharide (LPS) and LPS binding protein. Science 249:1431–1433
Xu SH, Inenaga K, Honda E, Yamashita H (2000) Glutamatergic synaptic inputs activate neurons in the subfornical organ through non-NMDA receptors. J Auton Nerv Syst 78:177–180
Xu Z, Herbert J (1996) Effects of unilateral or bilateral lesions within the anteroventral third ventricular region on c-fos expression induced by dehydration or angiotensin II in the supraoptic and paraventricular nuclei of the hypothalamus. Brain Res 713:36–43
Xu Z, Johnson AK (1998) Central renin injections: effects on drinking and expression of immediate early genes. Brain Res 782:24–35
Xu Z, Ross MG, Johnson AK (2001) Intracerebroventricular carbachol induces FOS immunoreactivity in lamina terminalis neurons projecting to the supraoptic nucleus. Brain Res 895:104–110
Yamaguchi K, Morimoto A, Murakami N (1993) Organum vasculosum laminae terminalis (OVLT) in rabbit and rat: topographic studies. J Comp Neurol 330:352–362
Yamaguchi K, Sieber NC (2000) The capillary of the organum vasculosum laminae terminalis (OVLT) in rabbits is more permeable to horseradish peroxidase (HRP) than that in rats. J Electron Microsc 49:783–791
Yamamoto T, Suzuki H, Uemura H (1997) Endothelin B receptor-like immunoreactivity is associated with LHRH- immunoreactive fibers in the rat hypothalamus. Neurosci Lett 223:117–120
Yang M, Card JP, Tirabassi RS, Miselis RR, Enquist LW (1999) Retrograde, transneuronal spread of pseudorabies virus in defined neuronal circuitry of the rat brain is facilitated by gE mutations that reduce virulence. J Virol 73:4350–4359
Zarbin MA, Innis RB, Wamsley JK, Snyder SH, Kuhar MJ (1983) Autoradiographic localization of cholecystokinin receptors in rodent brain. J Neurosci 3:877–906
Zardetto-Smith AM, Gray TS (1987) A direct neural projection from the nucleus of the solitary tract to the subfornical organ in the rat. Neurosci Lett 80:163–166
Zhang J, Rivest S (1999) Distribution, regulation and colocalization of the genes encoding the EP2- and EP4-PGE2 receptors in the rat brain and neuronal responses to systemic inflammation. Eur J Neurosci 11:2651–2568
Zhang J, Rivest S (2000) A functional analysis of EP4 receptor-expressing neurons in mediating the action of prostaglandin E2 within specific nuclei of the brain in response to circulating interleukin-1beta. J Neurochem 74:2134–2145
Zhang J, Rivest S (2001) Anti-inflammatory effects of prostaglandin E2 in the central nervous system in response to brain injury and circulating lipopolysaccharide. J Neurochem 76:855–864

Subject Index

Printing (Computer to Plate): Saladruck Berlin
Binding: Stürtz AG, Würzburg

GPSR Compliance

The European Union's (EU) General Product Safety Regulation (GPSR) is a set of rules that requires consumer products to be safe and our obligations to ensure this.

If you have any concerns about our products, you can contact us on ProductSafety@springernature.com

In case Publisher is established outside the EU, the EU authorized representative is:

Springer Nature Customer Service Center GmbH
Europaplatz 3
69115 Heidelberg, Germany

Batch number: 10704713

Printed by Printforce, the Netherlands